D0164724

Inorganic Chemistry in Biology

Patricia C. Wilkins

Retired, Previously Research Associate, Department of Biological Sciences, University of Warwick
and

Ralph G. Wilkins

Emeritus Professor of Chemistry, New Mexico State University

Series sponsor:  ZENECA

ZENECA is a major international company active in four main areas of business: Pharmaceuticals, Agrochemicals and Seeds, Specialty Chemicals, and Biological Products.

ZENECA's skill and innovative ideas in organic chemistry and bioscience create products and services which improve the world's health, nutrition, environment, and quality of life.

ZENECA is committed to the support of education in chemistry and chemical engineering.

OXFORD NEW YORK TOKYO
OXFORD UNIVERSITY PRESS
1997

QP
531
.W54
1997

Oxford University Press, Great Clarendon Street, Oxford OX2 6DP

Oxford   New York
Athens   Auckland   Bangkok   Bogota   Bombay   Buenos Aires
Calcutta   Cape Town   Dar es Salaam   Delhi   Florence   Hong Kong
Istanbul   Karachi   Kuala Lumpur   Madras   Madrid   Melbourne
Mexico City   Nairobi   Paris   Singapore   Taipei   Tokyo   Toronto
and associated companies in
Berlin   Ibadan

Oxford is a trade mark of Oxford University Press

Published in the United States
by Oxford University Press Inc., New York

© Patricia C. Wilkins and Ralph G. Wilkins, 1997

All rights reserved. No part of this publication may be
reproduced, stored in a retrieval system, or transmitted, in any
form or by any means, without the prior permission in writing of Oxford
University Press. Within the UK, exceptions are allowed in respect of any
fair dealing for the purpose of research or private study, or criticism or
review, as permitted under the Copyright, Designs and Patents Act, 1988, or
in the case of reprographic reproduction in accordance with the terms of
licences issued by the Copyright Licensing Agency. Enquiries concerning
reproduction outside those terms and in other countries should be sent to
the Rights Department, Oxford University Press, at the address above.

This book is sold subject to the condition that it shall not,
by way of trade or otherwise, be lent, re-sold, hired out, or otherwise
circulated without the publisher's prior consent in any form of binding
or cover other than that in which it is published and without a similar
condition including this condition being imposed
on the subsequent purchaser.

A catalogue record for this book is available from the British Library

Library of Congress Cataloging in Publication Data
(Data available)
ISBN   0 19 855933 X

Typeset by the authors
Printed in Great Britain by
The Bath Press, Bath

Series Editor's Foreword

Life on this planet is based on the raw materials available. A substantial proportion of the Periodic Table is utilized within biology, with the roles of the elements being governed by their availability and position in the Periodic Table. Here inorganic chemistry blurs into biology and environmental science, and for many students acquires new life.

Oxford Chemistry Primers are designed to give a concise introduction to all chemistry students by providing the material that would usually be covered in an 8–10 lecture course. As well as providing up-to-date information, this series also provides the explanations and rationales that are the framework of Inorganic Chemistry. Patricia and Ralph Wilkins have provided a fascinating and authoritative description of the place of inorganic chemicals in biology. It will either match specialist courses on this subject or serve as a companion to descriptive inorganic chemistry courses.

John Evans
Department of Chemistry, University of Southampton

Preface

The recent flood of bioinorganic chemistry texts attests to the vitality and significance of the subject. We thought that a primer could be very useful to senior undergraduates and postgraduate students, as well as being a refresher for research practitioners.

The principles of metallobiomolecule construction and behaviour are covered concisely in the first four chapters. Dave Fenton's book in this series, 'Biocoordination Chemistry', treats this aspect in greater detail and can serve as a complementary text. The bulk of the book is descriptive inorganic chemistry dealing with the import in biology of about thirty elements from the s-, d- and p-blocks of the periodic table. We have tried, where possible, to relate the chemistry to the environment, agricultural problems, clinical use and health and the often unique and fascinating behaviour of a range of critters. We decided, after much thought, to omit all references to individual scientists. Their names can be found in text books and the scientific literature.

John Evans, Mike Johnson and Lynn and Frank Guziec read the manuscript and made helpful comments. Steve Tate was a tower of strength with computer advice. The staff at OUP were extremely helpful during the various stages in the production of the book. All these people have our heartfelt thanks.

Finally, we dedicate this book to Sue, C.W., Keith, Dan, Linda and Kell with all our love.

Kenilworth
Warwickshire
August 1996

P. C. W.
R. G. W.

Contents

1 The important elements in biology

The importance of certain elements in biochemistry has long been appreciated. Only since the late 1960s however, has the interplay of inorganic chemistry and biochemistry been a separate discipline, signalled by the formation of interested groups of scientists, the holding of international meetings on the subject and the inclusion of the topic as a separate chapter in inorganic chemistry textbooks! The change of the early designation of the subject from inorganic biochemistry to biological inorganic (or bioinorganic) chemistry was an indication of the subtle change in emphasis. The subject is now a firmly entrenched and vibrant area of research. Solutions to the problems which the subject presents require the inputs of chemistry, physics and molecular biology, and the topic impacts on medicine, pharmacology, agriculture and environmental sciences. In recent years, the field has moved from one of characterization (although still important) to one of probing biological structure and function and the incursion of molecular biology. It is hoped that this book will give at least a taste of the rich diet afforded to practioners of the subject.

1.1 Essential elements

So far there are 112 known elements. All those above element 93, and three below (Tc, At and Fr), are artificially produced. About thirty elements are recognized as being indispensable to some form of life. Iron was known as an essential element as early as the seventeenth century. Most of the remainder were not diagnosed as vital until the latter half of this century and there is still some uncertainty about the necessity of some of the elements. Examination of Table 1.1 shows that it is the lighter elements which are those essential to life and which are also the abundant ones in the biosphere.

Q. What are the sources of the abundant elements in the biosphere?

Table 1.1   Abundant elements in the biosphere (shaded) and essential elements in living organisms (bold)

s-block		d-block										p-block				
1	2	3	4	5	6	7	8	9	10	11	12	13	14	15	16	17
H																
Li	Be											B	C	N	O	F
Na	Mg											Al	Si	P	S	Cl
K	Ca	Sc	Ti	V	Cr	Mn	Fe	Co	Ni	Cu	Zn	Ga	Ge	As	Se	Br
Rb	Sr	Y	Zr	Nb	Mo	Tc	Ru	Rh	Pd	Ag	Cd	In	Sn	Sb	Te	I
Cs	Ba	La	Hf	Ta	W	Re	Os	Ir	Pt	Au	Hg	Tl	Pb	Bi	Po	At

Further, there is even a striking parallelism between the concentrations of the dominant ions in seawater and, for example, in blood plasma, which suggests that our ancestry is rooted in the sea (Table 1.2).

Several ascidian species (e.g. sea-squirts) have the ability to concentrate the vanadium in sea-water (~ 10 nM V(V)) to as high as 0.15 M in their blood cells. In these, vanadium is probably present as a V(III) complex, but its structure and function are uncertain.

Table 1.2 Concentrations (mM) of some ions in seawater and blood plasma

Ion	Seawater	Blood plasma
Cl^-	550	100
Na^+	470	140
Mg^{2+}	50	2
K^+	10	10
Ca^{2+}	10	3
MoO_4^{2-}	0.2 μM	2–20 μM
First row transition elements	1–10 nM	2–20 μM

In both media of course the elements hydrogen and oxygen dominate. It is significant that one of the only two heavier transition metals which are essential (Mo) also appears in relatively large concentrations in seawater compared to all the other transition elements. The essential elements are derived from the s-, p- and d-blocks. The f-block (lanthanides and actinides) is not represented at all. The general functions of each block are different as detailed below.

Q.What criteria do you think might be used for an element to be considered essential?

s-block:

H		
Na		Mg
K		Ca

- most abundant metal ions in biology occurring in most cells at fairly high concentrations (~ mM)
- crucial for plant growth (K)
- difficult to monitor
- important skeletal role (Ca)
- trigger a wide range of biochemical processes (Ca, Mg)
- activators of enzyme action (K, Mg) and stabilizers of biomolecular structures (Mg, Ca)

d-block:

V	Cr	Mn	Fe	Co	Ni	Cu	Zn
	Mo						
	W						

- prevalent in biology, usually in trace amounts
- all found in humans in gram amounts (Mn, Fe, Zn) or less
- usually easily monitored
- important in metalloproteins, functioning in catalytic, structural and regulatory roles; control of gene activity (Zn)
- represented in all six enzyme classes; important in the activation of H_2, O_2, N_2 and CO_2
- participants in electron transfer, respiratory chain (Fe, Cu), photosynthesis (Mn, Fe, Cu) and O_2 storage and transport (Fe, Cu)

p-block:

Q. Why do you think the first row transition elements (rather than the second or third rows) are used almost exclusively?

B	C	N	O	F
	Si	P	S	Cl
		As	Se	Br
	Sn			I

- constituents of living matter (H_2O and organic compounds, carbohydrates, nucleic acids and proteins)
- present in important gaseous molecules
- as anions, aid in the assembly of skeletal material
- C, H, N and O comprise 99% of the human body (gram amounts)

The importance of organic compounds in the functioning of the human body is well known. Some idea of the vital role played by elements other than carbon for specific functions is shown in Table 1.3. We shall consider the entries in more detail throughout the book.

Table 1.3 The impact of inorganic species in the body

Body component or function	Element or compound
Teeth and bone construction	Ca, F, P
Urinary and renal stones	Ca
Transport and storage of O_2	Fe
Blood pressure and blood coagulation control	Na, Cl, NO, Ca
Muscle contraction	Ca, Mg
Respiration	Fe, Cu
Cell division	Ca, Fe, Co
Gut contraction and food movement; penis erection	NO, Ca
Control of pH in blood	CO_2, Zn
Thyroid function	I

Obviously, serious ailments and diseases can be caused in humans by an incorrect balance of elements. Some examples of these are shown in Table 1.4. It should be emphasized that there is always a fine balance between a deficiency and an excess intake (both of which could lead to death) and a short range of daily intake which is essential and optimal.

Cobalt occurs only in vitamin B_{12} in humans and must be supplied in the diet, since the body cannot synthesize the vitamin. The liver can store enough vitamin B_{12} for several years. Its absence can result in pernicious anemia. It has been reported that some heavy beer drinkers developed cardiomyopathy, which was attributed to the cobalt content of some beer added to enhance the head.

Table 1.4   Ailments and diseases caused by a deficiency (d) or excess (e) of essential elements

Ailment or disease	Element
Anemia	Fe (d) Co (d) Cu (d) Mo (e)
Lung diseases	Si (e) Ni (e) Cr (e)
Psychiatric disorders	Mn (e)
Goitre	I (d and e)
Heart failure	Co (e)
Convulsions	Mg (d)
Wilson's disease, Menke's syndrome	Cu (d and e)
Inhibited growth	Si, V, Ni, Zn, As, Mo and Mn (all d)

1.2 Non-essential but valuable elements

A few elements have become of marked clinical value. These are shown in Table 1.5 and are discussed further in Chapters 5 and 7.

Table 1.5   Clinical value of some elements

Value	Element
Psychotropic drug for manic-depressive illness	Li
Diagnostic imaging	Tc-99m
Magnetic resonance imaging enhancement	Gd
Cytotoxic drug	Pt
Antiarthritic drug	Au

1.3   Positively lethal elements

Many elements are highly toxic. Some which are likely to be encountered in present day society are shown in Table 1.6. The soft cations (heavy metal ions) bind sulfur ligands preferentially (Section 2.3) and are likely to bind S-containing amino acids such as cysteine and thus interfere with metabolic processes. Hard cations, e.g. Al^{3+}, will bind strongly to N- and O-donors (Section 2.3), i.e. to many biomolecules and thereby deactivate them. All toxic elements are chemically similar to an essential element and therefore interfere with the function of the latter. The two similar elements may be in the same group of the periodic table or have the same size ions. Many radioactive compounds, in particular those of Sr-90 and Pu-239, are highly toxic. Treatments for the removal of toxic cations are discussed in Section 7.8. They require the judicious use of strong chelating agents.

Table 1.6   Toxic elements

Aluminium

Highly toxic. Al^{3+} causes anemia (can replace Fe^{3+}), dementia and death. Can replace Mg^{2+}, since it binds more strongly to the same ligands. Lowering of pH in soils increases bioavailability of Al(III).

Arsenic

Powerful poison. As(V) simulates P(V). As(III) reacts with thiol groups thus affecting enzymes. Complete absence in diet leads to growth and reproductive disorders. Occurs naturally as non-toxic $Me_3As^+CH_2CO_2^-$ in fish and lobsters.

Beryllium

Most toxic of all elements. Be^{2+} can interfere with many of the functions of Mg^{2+}. Ingestion of beryllium compounds causes lung cancer and other pulmonary diseases.

Cadmium

Highly toxic leading to dysfunctioning of the kidneys. Cd^{2+} can replace Zn^{2+} in S-containing enzymes and substitute for Ca^{2+} in bone. Biological retention of many years. Organisms may react to Cd^{2+} (and other heavy metal ions) by increasing production of metallothioneins which act as clearing agents.

Mercury

Locates in liver, kidney and brain ('mad-hatter', in the last century hatters used $Hg(NO_3)_2$ for cleaning felt hats). Hg is less toxic than Hg^{2+} and RHg^+ ions, which are soluble and move easily through membranes (e.g. blood–brain). Bacterial resistance to Hg(II), and associated metal-regulated transcription and detoxification have been well studied (MerR, Section 7.6).

Lead

Long history of toxicity. Pb^{2+} causes anemia by inhibiting heme synthetase which catalyses Fe incorporation into porphyrins. Organometallic Pb compounds cross the blood–brain barrier and attack the central nervous system.

Selenium

Toxicity leading to hair and nail loss, but component of glutathione peroxidase which protects red blood cells from H_2O_2. Essential nutrient in the soil. Some plants can remove Se salts and transform them into the much less toxic gas, Me_2Se.

Strontium-90

Long term effect on bone marrow function. Sr^{2+} ends up in mineralized portion of bone (simulating Ca^{2+}). Long biological retention. Removal promoted by Ca^{2+} complexing ligands (e.g. EDTA, Section 7.8).

Thallium

Neurotoxin. Tl^+ stable ion, radius similar to that of K^+ and can replace K^+ in many of its functions.

Cadmium in the soil from a zinc mine led to severe contamination of vegetables and fruits in a Somerset village in the late 1970s. More seriously, about 100 deaths were caused by Cd contamination of rice in Toyama, Japan in the 1950s. *Itai–itai* (ouch–ouch) disease causes demineralization of bones which then shrink.

2 The composition of metallobiomolecules

In this chapter we are concerned with the make-up of the metallobiomolecule, and in particular with that portion in which the metal ion resides. The rules which govern the structures of biomolecules are much the same with or without a metal ion, although there can be significant perturbations due to the presence of the metal. The metal–ligand bond is central to many aspects of the behaviour of metallobiomolecules. In nearly all metalloproteins and metal complexes of biologically important materials (carbohydrates, nucleic acids etc.), the metal ion will be attached to the ligands by nitrogen, oxygen or sulfur donor atoms. The number of donor atoms bound to the metal ion constitutes the *coordination number,* which is often 4, 5 or 6, but can vary from 3 to even 8. The geometries observed are likely to be distorted from those normally encountered in simple metal complexes (Fig. 2.1) because of the constraints imposed by the protein structure and also to aid in the biological function of the metalloprotein.

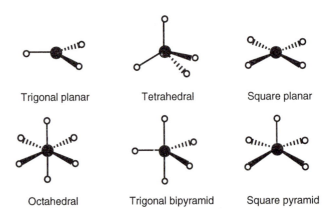

| Trigonal planar | Tetrahedral | Square planar |

| Octahedral | Trigonal bipyramid | Square pyramid |

Fig. 2.1 Coordination shapes shown by simple metal complexes. The square planar geometry is rarely exhibited by metal sites in biomolecules.

2.1 Biologically important ligands

A ligand may provide one (unidentate), two (bidentate) or more (multidentate) donor atoms to a metal ion. The most important of a number of uni- or bidentate ligands are H_2O, PO_4^{3-} (especially in nucleotides and nucleic acids), metalloenzyme substrates and inhibitors, small cellular constituents and certain organic cofactors. The two most prevalent *types* of ligands are (a) the side chains of amino acids which are constituents of the protein backbone and (b) macrocycles, exemplified by the porphyrin ring.

Table 2.1   Common amino acid chains as ligands in metalloproteins

R	Occurrence
CH_2–imidazole (his)	Predominant; usually attached via either N of imidazole ring
$CH_2CO_2^-$ (asp) and $(CH_2)_2CO_2^-$ (glu)	Useful bridging groups, binding two metals by carboxylate oxygens; also uni- and bidentate to a single metal
CH_2S^- (cys) and $(CH_2)_2S(CH_3)$ (met)	M–S linkage often associated with M = Fe^{3+}, Cu^{2+} and Zn^{2+}
$CH_2C_6H_4O^-$ (tyr)	M–O linkage often coordinated to M = Mn^{2+}, Fe^{3+} and Cu^{2+}

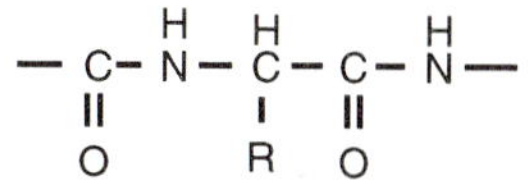

(a) Only eleven of the natural amino acids have sidechains which contain donor groups in R (see diagram) and just six of these are commonly observed in metalloproteins, Table 2.1. Several of the sidechains will coordinate to a metal ion to form chelate rings. There are several examples in Section 2.2 and throughout the book. In rarer cases the peptide carbonyl oxygen and the deprotonated peptide N^- can also be donor atoms.

(b) In the basic porphine (more commonly termed porphyrin) structure (Fig. 2.2) the central NH hydrogens are replaced by a metal ion, M^{n+}, which may locate within or slightly out of the plane described by the four heterocyclic nitrogens and M will usually represent iron.

Q. What are the various modes of coordination of the imidazole ring in histidine and the carboxylate group in aspartate or glutamate to a metal ion when these amino acids are part of a peptide chain?

Q. A number of isomers of porphyrin have been synthesized in the past few years (porphycene, N-confused porphyrin and others). What might their structures be and are they likely to coordinate to metal ions?

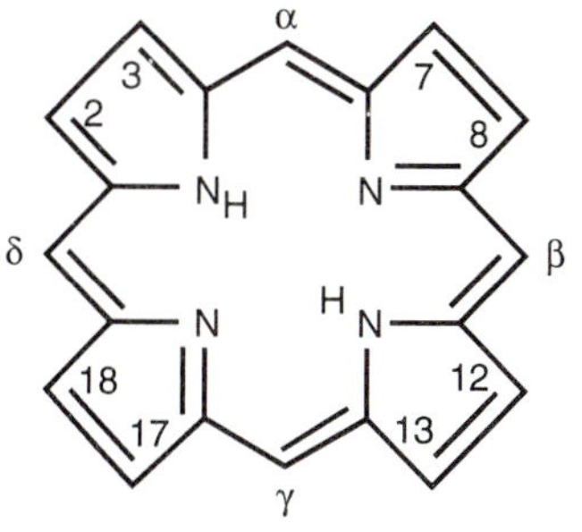

Fig. 2.2 Porphyrin structure with International Union of Biochemistry suggested numbering.

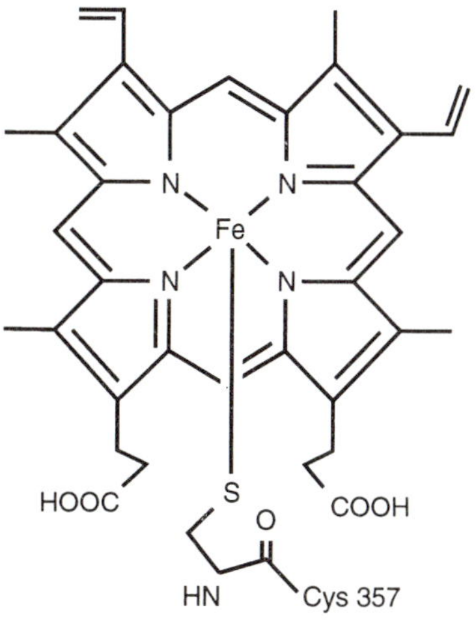

Fig. 2.3 Protoporphyrin IX in cytochrome P-450. The macrocycle is attached indirectly to the polypeptide chain via cysteine coordinated to the Fe. The sixth ligand may be H_2O.

We shall encounter a number of derivatives of the basic porphyrin structure with different peripheral groups and with different degrees of saturation of the rings (Table 6.1). A particularly important porphyrin is protoporphyrin IX, the iron substituted form of which occurs extensively in cytochromes, metalloenzymes and transport proteins (see Fig. 2.3). Other biologically important macrocyclic ligands are ionophores (Section 5.3) and siderophores (Section 7.5).

Each nucleotide segment (as in ATP, DNA and RNA) contains three potential donor sites for binding to metal ions (Fig. 2.4). These are:

(a) The nitrogen base, favoured by heavy metals. In nucleic acids, this bond is sometimes reinforced by additional binding to an adjacent base on the same or the complementary strand (in DNA).

(b) Sugar hydroxyl groups. A rare example is in the binding with Os(VI).

(c) Sugar bound phosphate oxygen atoms. This is a very important ligand site for electrostatic binding to hard metal ions such as Mg^{2+} (see Section 2.3).

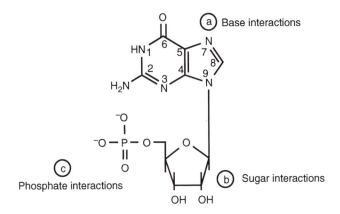

Fig. 2.4 Potential coordination sites in a nucleotide, guanosine 5'-monophosphate, GMP. Guanine N7 is often favoured for binding, since other base sites are less nucleophilic and, when part of a DNA strand, less accessible to the metal.

2.2 Metal sites

Families of metalloproteins often show topological similarities. The metzincins (Section 3.1) have 5-stranded β-sheets and 3-α-helices in typical sequential order. The 'cupredoxin fold' turns up in a number of type1 blue copper proteins (Section 4.2) and the 'four helix bundle' originally observed in hemerythrin (Section 8.7) has been found in many different proteins.

Many metalloproteins contain only one metal ion which is quite often part of the reaction site. For this reason alone we will tend to focus on that portion of the protein. We should not however lose sight of the fact that, as well as the coordinated ligands, a number of amino acid residues are nearby, or even distant, which play a vital role in the action of the protein (Fig. 2.5). Metal sites are often at the bottom of a cleft in the protein which allows access to substrate or inhibitor molecules. Proteins containing binuclear metal sites are also common, and in these the metal ions can be the same or different.

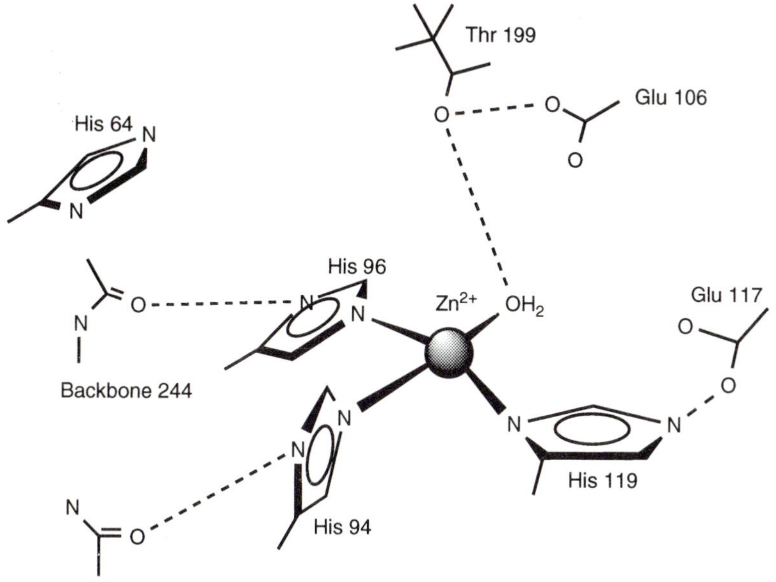

Fig. 2.5 Environment of zinc in carbonic anhydrase. The dotted lines indicate hydrogen bonds and the five-membered rings are imidazoles, with N coordinated. There is probably a fifth ligand (H_2O) associated with the Zn.

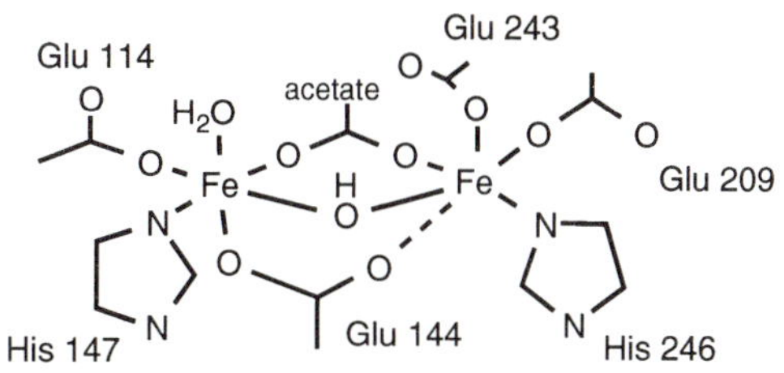

Fig. 2.6 Binuclear iron site in the hydroxylase protein of soluble methane monooxygenase (*Methylococcus capsulatus* (Bath)).

A similar site with two octahedral irons and bridging ligands including either O^{2-} or OH^- and asp or glu features in a number of non-heme iron proteins including the respiratory protein hemerythrin (Section 8.7), the R2 subunit of *E. coli* ribonucleotide reductase and purple acid phosphatase from bovine spleen (Section 7.2).

A binuclear iron site with bridging ligands (Fig. 2.6) appears in a number of proteins which have quite dissimilar functions, emphasizing the ability of Nature, after much trial and error, to finally select and stick by a useful type of structure. It is not surprising, when one considers the complicated and diverse functions of certain metalloproteins, to find that some of them require three or more metal ions and we shall encounter a number of these clusters later. The metal ions in these cases can have one or more roles, which include structural, catalytic or transport and storage. The cluster might even serve a function *in toto* as in the [4Fe–4S] unit in the electron carrying ferredoxins (Fig. 2.7) and more than one such cluster may be present in a protein.

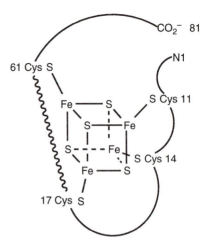

Fig. 2.7 The [4Fe–4S] cluster in *Bacillus thermoproteolyticus* ferredoxin. Each iron in the distorted cube is attached tetrahedrally to three S^{2-} ions and one cys residue from the protein. The size of the cube is greatly exaggerated. It is located near the surface of the protein.

Positively charged metal ions embedded deep in proteins must be neutralized by negatively charged ligands. These would be S^{2-} and cysS⁻.

2.3 Favoured donor atoms for metal ions

Cations which are small and highly electropositive (including H⁺) or which have a high charge are termed *hard metal ions*. Large cations, usually with a low charge, are referred to as *soft*. With regard to ligands, those which are very basic, i.e. have a high proton affinity (first row of the p-block) are termed *hard*, whereas other ligands are referred to as *soft*. In a large number of simple metal complexes formed from a cation (often called a Lewis acid) and a ligand (a Lewis base) it has been found that hard cations have a preference for hard ligands and that soft cations favour soft ligands. These generalizations apply quite well to metallobiomolecules also. Tables 2.2 and 2.3 clarify these ideas.

Table 2.2 Favoured combinations for biologically important cations and donor centres

Type	Cations	Donor centres
Hard	H⁺, Na⁺, K⁺, Mg^{2+}, Ca^{2+}, Mn^{2+}, Mn^{3+}, Fe^{3+}	Oxygen in H_2O, OH⁻, OR⁻, O^{2-}, PO_4^{3-}, NO_3^-, CO_3^{2-}, RCO_2^- (so includes side chains of glu, asp, tyr, ser and thr), –C=O (peptide), F⁻, Cl⁻ and NH_3
Soft	Cu⁺, Ag⁺, Pt^{2+}, Cd^{2+}, Hg⁺, Hg^{2+}	CN⁻, CO, S^{2-}, RSH and R_2S (so includes cys and met), I⁻
Borderline	Fe^{2+}, Co^{2+}, Ni^{2+}, Cu^{2+}, Zn^{2+}	N, O and S donors all possible, –C(=O)N– of peptide

Table 2.3   Examples of favoured combinations

Cation	Donor centres	Example
Ca^{2+}	Only O donors known so far	Numerous calcium binding proteins.
Mg^{2+}	O strongly preferred	Phosphate groups in DNA, RNA.
Mn^{2+}	Prefers O over N, S (can therefore be interchanged with Mg^{2+})	Xylose isomerase contains a variety of carboxylate bindings to Mn^{2+}.
Hg^{2+}	Favours S donors	MerR (metal regulatory protein that mediates metal responsive gene regulation). Mercuric reductase.
Zn^{2+}	N, O and S donors all encountered (favours his, glu, asp and cys)	Thermolysin and Zn-dependent endopeptidases. Metallothionein.
Cu^{2+}	N, O and S donors all encountered	Azurin contains 2his, cys, met and a $-C=O$ as Cu ligands.

Hard ligands stabilize the higher oxidation state(s) of the metal ion, whereas soft ligands prefer the lower oxidation state(s). The Fe(IV)=O (in P-450) and Co(I)–C (in coenzyme B_{12}) moieties are examples of this.

2.4   Types of bonding

Biomolecules are constructed using all types of bonding. Covalent and ionic bonds (and intermediate gradations) are strong and form the scaffolding of

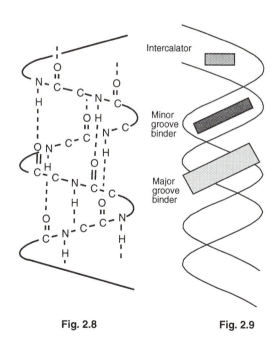

Fig. 2.8

Fig. 2.9

Fig. 2.8   Hydrogen bonding in the α-helix. Nitrogen and oxygen atoms from the main chain are H-bonded to each other within the α-helices. There are 3.6 residues (5.4 Å) per turn.

Fig. 2.9   Modes of interactions of molecules with DNA. The major groove is ~ 24 Å wide and deep whereas the minor groove is narrow (~10 Å) and shallow. The former is probably the main site for recognition by promoters and repressors.

biomolecules. Hardly less important are the much weaker hydrogen bonds (Fig. 2.8) and van der Waals interactions (Fig. 2.9) which provide the necessary subtlety to these structures. The types of bonds are compiled in Table 2.4. All of these types of bonding can contribute to the stability of a biological system. For example, in the construction of a membrane (Fig. 5.1) hydrophobic interactions and van der Waals attractive forces operate between the tails of the lipid bilayers and electrostatic and hydrogen bonding interactions occur between the polar heads and solvent water. Finally the individual molecules are held together by covalent bonds.

Q. Albumin is a major component of serum (the clear fluid which remains after blood clots) and of egg white. It aids in the transport, distribution and metabolism of endogenous and exogenous materials, including metal ions. In human serum albumin the terminal peptides are:

What groups are the likely binding sites for Cu^{2+}?

Table 2.4   Types of bonding

Bond type	Characteristics	Occurence
Covalent	Usually requires enzymatic action to rupture.	Within polypeptide and polynucleotide chains C–C, C–N, P–O etc. Ligand (substrate, inhibitor) binding to metal ion.
Ionic (also termed electrostatic or salt bridge)	Association of positively and negatively charged species.	Salt bridges, e.g. (asp) $CH_2CO_2^-$ ····$H_2^+N{=}C{=}C$ (arg) between subunits in deoxyhemoglobin. Association of hard cations and hard ligands (Mg^{2+}····OPO_3^{3-}).
Hydrogen	H atom shared between two other highly electronegative atoms (usually N and/or O). Relatively weak (~ 20 kJ mol^{-1} and easily formed and broken.	In α–helices (Fig. 2.8) and β-sheets of proteins. Base–base interactions on opposite strands of DNA double helix. Active sites of metalloenzymes (see Fig. 2.5).
Van der Waals (dipole – dipole)	Large number of atoms reinforcing transient weak polarizing effects (~4 kJ mol^{-1} for each pair).	Attractive forces at short distances between permanent dipoles, e.g. $-C^{\delta=}{=}O^{\delta-}$····$-C^{\delta+}{=}O^{\delta-}$ or induced dipoles, e.g. $C^{\delta+}-H^{\delta-}$. Intercalation of planar moiety between DNA base pairs (Fig. 2.9).
Hydrophobic	Intermolecular aggregation of non-polar species in an aqueous medium. Weaker than H-bonds.	Oil droplet formation. Folding of macromolecules. Formation of membranes. Association of small molecules with DNA in the minor groove (Fig. 2.9).

3 The examination of the properties of metalloproteins

A thorough examination of the physical and chemical properties of most metalloproteins will yield information on their overall structure but more especially insight into details of the active site. Since the latter will quite often encompass the metal, a good deal of attention will focus on the metal ion and its immediate environs.

3.1 General features

The presence of a metal ion in a biomolecule has not always been easy to detect. It was 50 years after the the discovery of urease (the first protein to be crystallized) and sometime after the isolation of certain hydrogenases, that nickel was found to be present in both. The iron content, in particular, of a number of metalloproteins (several binuclear iron proteins, Fig. 2.6 and iron–sulfur proteins, Fig. 2.7) had been wrongly assessed at first (invariably underestimated), resulting in serious misjudgments of the number of irons present in the protein. Metal analyses of cytochrome c (cyt c) oxidase (Table 6.3) indicate that in the purest preparations with the highest activity, the Cu/Fe ratio is 3:2 and not 2:2 as had been reported. This has been confirmed by a recent X-ray crystallographic study. The corrections have been made, largely because of the availability of improved analytical methods and a battery of spectroscopic techniques (Section 3.2).

The methods for the determination of the basic characteristics of a metalloprotein are those which are applicable to all proteins (molecular weight, amino acid sequence etc.). With respect to the assessment of molecular weights, there have been exciting developments in mass spectrometry techniques since the 1980s. Fast atom bombardment, matrix-assisted laser desorption ionization and electrospray ionization allow for accurate molecular weight determinations. In addition, a combination of fast atom bombardment and tandem mass spectrometry can yield sequence details for polypeptides, polynucleotides and even for proteins of molecular weights from 10–15 kDa when these are available in only nanomole amounts.

Motifs

The comparison of the amino acid sequences of a metalloprotein from a number of sources or of different metalloproteins with a common function can, in favourable cases, allow reasonable guesses as to the amino acids which are essential for its action. These will be invariant amino acids and will almost certainly include the ligands associated with the metal. The alignment of the primary sequences of the R2 subunit of ribonucleotide

Motifs abound. Thus, although the human matrix metalloproteinases and other zinc proteinases from bacteria, crayfish and snake venom have large overall differences, they all contain the conserved sequence: his*-glu-xx-his*-xx-gly-xx-his*, with the three his* coordinated to the catalytic Zn (metzincins). Part of this consensus is also contained in the zinc enzymes thermolysin, angiotensin-converting enzyme and enkephalinase (Table 7.4). Predictions of the amino acids adjacent to the metal site from sequence examinations of zinc enzymes, have been confirmed by NMR and X-ray analyses.

reductase and the α-subunit of the soluble methane monooxygenase hydroxylase, using the amino acids known to be iron ligands in R2 led to a prediction of the active site structure of the latter, which was later confirmed by X-ray crystallography (Fig. 2.6).

3.2 Characterization

Several properties are important in shedding light on the nature of the metal site. Paramount are the wealth of spectroscopic properties which are usually available.

Spectroscopic properties

Spectroscopic techniques are extremely important in the study of metalloproteins. A short description and the relevance of such techniques is presented in Table 3.1. More detailed explanations of the methods can be obtained from the 'Further reading' section. Metalloproteins usually exhibit a number of spectroscopic features. The more common techniques (e.g., electronic and vibrational spectroscopies, CD and EPR) were used in early studies for categorizing and probing the structure at the metal site. For example, these spectroscopic methods applied to plastocyanin led to the correct prediction of met, cys and 2 his as the Cu ligands arranged in a distorted tetrahedron. In addition, monitoring the interactions of metalloproteins with their biological reaction partners (substrate, inhibitor etc.) usually requires a spectroscopic probe and UV-vis, fluorescence, resonance Raman, EPR and NMR are particularly useful. A wide range of spectroscopic techniques are now available to the bioinorganic chemist (Table 3.1) and their value has been enhanced in recent years by the acquisition of computer controlled instruments and Fourier-transform methodologies. The techniques can help, in varying degrees, in the determination of the:

(a)   number and nature of the ligands attached to the metal;
(b)   coordination geometry around the metal;
(c)   oxidation and spin state of the metal;
(d)   characterization of metal clusters;
        and, with only the most sophisticated approaches
(e)   the metal–ligand and metal–metal distances and angles.

Characterization is also more likely if more than one technique is applied. Comparison of the properties of the metal site in a protein with those of simple metal complexes, of known structure, has also proved to be very valuable (Section 3.5). For some of the techniques to be effective, it is necessary to replace a spectroscopically 'silent' native metal ion (e.g. Zn^{2+}) by an apt one (e.g. Co^{2+} for UV-vis or $^{113}Cd^{2+}$ for NMR, see Table 3.3).

In general, the techniques shown in Table 3.1 are of increasing diagnostic value but are also of increasing complexity and cost and therefore decreasing availability! Synchrotron radiation sources have made EXAFS a viable technique and have improved the capabilities of X-ray crystallography. These methods, based on the absorption and diffraction of X-rays, are among the most powerful and definitive ones for a complete structure determination. It is also now possible to determine the structures of medium sized metalloproteins (and non-metalloproteins of course) *in solution* by sophisticated NMR techniques. MCD, Mössbauer, EPR and EXAFS experiments are usually performed at low temperatures.

Table 3.1   Spectroscopic properties and their value in characterizing the metal site

Method	Information obtained	Value and application (Section reference)
Ultraviolet (UV) and visible (vis) absorption	Energies and intensities of d→d, ligand-to-metal charge transfer (LMCT) and $\pi \rightarrow \pi^*$ transitions.	Readily available technique. Fe(III), Cu(II) proteins (LMCT) and metalloporphyrins ($\pi \rightarrow \pi^*$) are intensely coloured (3.5, 4.2, 6.3, 8.2).
Raman and resonance Raman, scattered light intensity	Vibrational modes at chromophoric site.	More sensitive than infrared with less interference from H_2O. Fe , Cu proteins (3.5, 8.2).
Circular dichroism(CD) and magnetic circular dichroism (MCD)	Ellipticity of chromophore (CD) or magnetically induced chromophore (MCD).	Protein and nucleic acid structures. MCD determines magnetic coupling and magnetic susceptibility. Fe proteins (6.2).
Mössbauer	Isomer shift (δ) and quadrupole splitting (ΔE_Q).	Oxidation, spin states and magnetic coupling in iron clusters (6.2, 8.3).
Electron paramagnetic resonance (EPR)	g values and hyperfine coupling constants. Interaction between spin of unpaired electron and nuclei at or near the metal site.	Number and symmetry of unpaired electron(s) at site. Detection of radicals (6.2), FeS clusters (6.2), Cu(II) proteins (4.2), hydrogenases (8.3).
Nuclear magnetic resonance (NMR)	Chemical shift (δ). ^{1}H, ^{13}C, ^{15}N and ^{31}P particularly useful nuclei.	Environment of paramagnetic metal site. Complete structure of small proteins in solution, e.g. metallothioneins (3.4, 7.5). Interactions of Pt(II) drugs with DNA (7.8).
Extended X-ray absorption fine structure (EXAFS)	Electronic transitions of core s-electron from metal.	Metal–ligand (± 0.02 Å) and metal–metal distances, but not bond angles. Rubredoxin (6.2), metallothioneins (7.5).
X-Ray single crystal diffraction (X-ray crystallography)	Fourier transform of crystal diffraction data using monochromatic X-rays (Bragg) or a broad spectrum of X-rays (Laue) produced by a synchrotron.	Complete solid state protein structure. Structures of proteins and nucleic acids reported weekly. Numerous examples (6.2, 7.5, 8.3).

Magnetic properties

Several spectroscopic techniques (e.g. MCD, Mössbauer, EPR and NMR) which focus on the metal ion can be used to determine the magnetic susceptibility and thus the magnetic moment and spin state at the metal centre. These parameters are difficult to obtain by the conventional methods used for metal complexes because of the very large diamagnetic background of the protein and water. This problem can be overcome by the use of a superconducting quantum interference device (SQUID) which accurately measures very low level magnetic fluxes and has been used to observe where and how signals travel in the brain. Low temperatures are required and the technique is very complicated, however.

Redox potentials

The reduction potential, $E°$ of the couple in eqn 3.1 is related to the measured

$$M^{m+} + e^- \rightleftharpoons M^{(m-1)+} \tag{3.1}$$

potential, E, and the concentrations of the oxidized, [oxid], and reduced, [red], forms by the Nernst equation:

$$E = E° + (RT/F) \log [red]/[oxid] \tag{3.2}$$

If a series of spectra of a redox-active metalloprotein are obtained at different applied potentials (E), these can be used to calculate the [red]/[oxid] ratio and from eqn 3.2 the value of $E°$, the standard reduction potential, see Section 6.4, can be obtained. A small organic molecule (mediator) is usually added to establish the redox equilibrium between the electrode and the metalloprotein. No direct contact of the protein with the electrode is then necessary. This is not the case with cyclic voltammetry (CV). Here variations of cathodic and anodic currents as the electrode potential is changed are recorded, allowing the determination of $E°$ values. There are formidable problems if the redox centre is buried in a large protein and is inaccessible to the electrode. Modified gold or pyrolytic graphite electrodes are occasionally successful. Cytochrome c shows well behaved cyclic voltammograms at a gold electrode and its interactions with non-redox partners, e.g. cytochrome b_5 or plastocyanin, have been studied using CV.

Q. How might one determine the binding constant of Co(phen)$_3^{3+}$ with DNA by cyclic voltammetry?

The term *labile* is used to denote a system (i.e. ligand binding) which occurs within seconds or faster. *Inert* systems are those which only slowly equilibrate.

3.3 Reaction rates

There is a wide range of rates associated with the reactions of metalloproteins. The first visible events when oxygen binds to the iron in myoglobin (Section 8.7) or after the photosynthetic reaction centre is irradiated (Section 6.4), take place in very short times, 10^{-12} s (ps). Many other important steps in bioinorganic reaction cycles are also rapid including acid–base reactions, electron transfer processes and spin-state changes, protein–protein interactions, protein folding and interactions of ligands with metalloproteins (a vast area including metalloenzyme catalysis) all of which often occur within 10^{-6} s (μs) or 10^{-9} s (ns). The fastest second-order reactions are diffusion-controlled, with rate constants ~10^8–10^9 M^{-1} s^{-1}. Some catalysis by metalloenzymes or conformational changes in metalloproteins can, on the other hand, be quite slow, may take seconds or minutes for completion and be easily measurable by conventional methods. There are a variety of techniques now available for the study of rapid reactions (Table 3.2).

Table 3.2   Techniques for the study of rapid reactions

Technique	Time (s)^a	Comments
Stopped-flow. Rapidly mixed solutions abruptly stopped and analysed near mixer.	10^{-3} (10^{-5})	Most popular general method. There are a number of commercial instruments available with a variety of monitoring methods.
Temperature-jump relaxation.	10^{-6} (10^{-9})	By-passes mixing time limitations.
NMR.	10^{-5} (10^{-7})	Fast exchange processes by NMR line broadening.
Pulse radiolysis. Generation of highly reactive radicals by ionization of H$_2$O (eqn 8.14).	10^{-8} (10^{-12})	Reactions of e$^-$, OH, H, HO$_2^-$ and other radicals with substrates can be examined.
Laser photolysis. Short burst of high intensity monochromatic light.	10^{-8} (10^{-13})	Produces highly reactive state in times as short as ps.

^{a} Usual and extreme (parenthesis) times of reaction.

Monitoring a reaction by more than one technique is often rewarding, since each usually provides different information. A good example is the study of the folding of a protein *in vitro*. Stopped-flow with fluorescence monitoring (environment of trp and tyr residues) and far- and near- UV CD (formation of different secondary structure) give insight into the complex pathways which attend change from a denatured to a native structure, which is often complete within seconds.

3.4   Altered metalloproteins

Two significant changes to metalloproteins can be made which involve the metal and the environment around the metal. These changes may help in the characterization of the metal site (Section 3.2), in understanding the function of the metalloprotein and in delineating the roles of each metal ion in multimetal-containing proteins.

Tinkering with the metal

A metal ion can generally be removed by gentle dialysis of the metalloprotein against a solution containing a chelating ligand, often at pH 5.5–7.5. The equilibria are continuously shifted from left to right in eqns 3.3 and 3.4,

$$M^{n+} \text{(protein)} \rightleftharpoons M^{n+} + \text{apoprotein} \qquad (3.3)$$

$$M^{n+} + \text{chelate} \rightleftharpoons M^{n+}\text{(chelate)} \qquad (3.4)$$

leaving demetalloprotein (apoprotein) inside the dialysis tubing. Apoprotein is usually devoid of any activity (e.g. Zn and Fe enzymes) or is less stable (Ca binding proteins). Treatment of the apoprotein with metal ions in solution produces a metalloprotein (holoprotein). If the native metal ion is used, the original metalloprotein with full function can usually be regenerated. Other metalloderivatives are generally much less active than the native form, or completely inactive. In cases where both have been examined by X-ray crystallography, there is usually little difference in the structures of the apo and native metalloproteins (see however transferrin, Section 7.5). This is interesting in, for example, plastocyanin, since it shows that the unusual geometry around the copper (Section 3.2) is imposed by the protein. Zinc proteins, although extremely important, are particularly difficult to study as they have few spectroscopic features. Replacement of Zn^{2+} by Co^{2+} and by Cd^{2+} can give clues as to the role of the metal ion (Table 3.3).

Q. What other methods might be used to prepare metalloderivatives?

The folding of an apoprotein from denatured to a native structure can usually occur without the help of a metal ion. Exceptions are metallothioneins and zinc fingers.

Q. What alternative metal ions might be useful for 'visualizing' Ca^{2+} and Mg^{2+} containing proteins?

Q. Yeast enolase requires Mg^{2+} (in an unusual five-coordinate geometry). If Mg^{2+} is replaced by Ca^{2+} inactivity results, but if the replacement is by Zn^{2+}, 80% of activity remains. Why do you think this is so?

Table 3.3 Examples of apoprotein preparation and value

Protein	Metal replacement and method	Parameter highlighted	Value
Carbonic anhydrase	Zn(II) removed with 1,10-phenanthroline. Co(II) added, 50% activity recovered.	d-d spectra	Indicated five-coordinate Co(II) in acid and tetrahedral Co(II) in basic (active) form.
Metallothioneins (Zn_nCd_{7-n} from liver tissue)	Zn(II) and Cd(II) removed by lowering the pH to ~2. Added ^{113}Cd(II) gives ^{113}Cd$_7$ metallothionein.	^{113}Cd NMR	Complete 3D structure in mM solutions (one of the first metalloproteins thus examined, Fig. 7.6).

A minority of metalloproteins have more than one polypeptide chain. Hemoglobin has four, designated α_1, α_2, β_1 and β_2. Each subunit binds an Fe protoporphyrin IX centre and has a three-dimensional structure similar to that of myoglobin. The four chains are held together in a compact structure. At the interfaces, alteration of amino acids using molecular biology, can change the characteristics of HbA into those, for example, of trout (Section 8.7), crocodile (Fig. 8.6) or bar-headed geese. These last contain hemoglobin with a high O_2 affinity, which is necessary for them to migrate across the Himalayas.

Changing selected amino acids

Altering amino acid residues at or even distant from the active site of a protein can modify the normal function of that protein. There are over 400 variants of human hemoglobin, in most of which a single amino acid is different from that in 'normal' hemoglobin (HbA). In a few cases this has devastating consequences. Sickle-cell anemia was the first human disease to be traced to an abnormal protein, namely hemoglobin S (HbS). In it, a glutamate residue (glu 6) is replaced by valine in the β-subunits of hemoglobin. This creates a hydrophobic patch on the surfaces of the β_2-subunits, and one of the patches is able therefore to fit into the hydrophobic pocket of another β_1-subunit. This leads to the production of fibres (by polymerization) in deoxyHbS and the 'sickling' of the red blood cells which reduces their ability to transport O_2 (Section 8.7).

It is now possible (routinely) to replace one amino acid by another using genetic engineering techniques. In this method, the gene encoding the protein is altered and the recombinant protein is expressed in a suitable host (site directed mutagenesis). For example, endopeptidase (NEP) is a membrane-bound zinc metalloenzyme. Replacement of glutamic acid by valine at position 646 (E646V) markedly reduces the catalytic (hydrolytic) activity of NEP. This effect, coupled with other data, strongly indicated that glu 646 is one of the ligands to zinc (his 583 and his 587 are others). In the mutant enzyme, the active site has been disturbed by modifying the coordination around the zinc.

Site directed mutagenesis has been used very effectively to build metal coordinating ability into a protein. Histidine or cysteine residues have been introduced into specific positions in certain proteins. Rat anionic trypsin catalyses the hydrolysis of peptide bonds on the carboxyl side of lys or arg residues. This trypsin is a member of the serine protease family, most of which contain a serine, aspartate and histidine triad at the active site (Fig. 3.1). Replacement of arg 96 by his in rat anionic trypsin (R96H) does not change the activity of the enzyme *until metal ions are added*. In the recombinant enzyme his 57 and the engineered his 96 can form a chelate with added metal ion, but only after rotation of his 57 around a C_α–C_β bond thereby removing his 57 from the active site (Fig. 3.1). Copper (II) ion forms the most stable chelate (confirmed by a crystal structure) and is therefore a potent non-competitive inhibitor ($K_i = 21\ \mu M$) of the enzyme. Removal of the Cu^{2+} restores the activity.

For the interaction of enzyme (E) with inhibitor (I), $E + I \rightleftharpoons EI$, $K = K_i^{-1} = $ [EI]/[E][I]. Chemists tend to use binding constants, i.e. K, while biochemists favour the dissociation constant, i.e. K_i. The combination of two molecules is a widespread occurence. It can involve interaction of O_2 or CO with a respiratory protein (Section 8.7), substrate or inhibitor binding to an enzyme or protein–protein association.

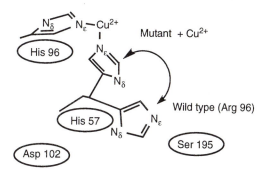

Fig. 3.1 Creation of chelate site and its influence on rat anionic trypsin activity. The constellation comprising his 57, asp 102 and ser 195 is referred to as the catalytic triad. Serine carries out nucleophilic attack on the peptide bond. Histidine takes up H^+, aided by the carboxylate of aspartate. Trypsins are members of the serine protease family which are widespread and occur as the digestive enzymes in the pancreatic juice of mammals.

Q. What would you expect the relative effects of Cu^{2+}, Ni^{2+} and Zn^{2+} to be on the activity of the mutant trypsin (R96H)?

Site directed mutagenesis is proving to be a valuable tool in examining the mechanisms of all types of enzyme catalysed reactions; the paths of electron transfer in proteins; the interactions between proteins and the details of metal sites in proteins. Examples of all of these will occur throughout the book.

Q. How might the ability to bind metal ions by recombinant proteins be used in protein purification?

Q. What can you see as potential disadvantages of metal ion replacement and genetic engineering in studying metalloproteins?

The electron spin due to each Cu(II) (d^9) may be cancelled by antiferromagnetic coupling. The net result is no unpaired electrons and no EPR signal nor magnetic splitting in the Mössbauer spectrum. The important parameter is $-J$ (cm^{-1}) which is a measure of the extent of magnetic exchange in antiferromagnetic coupling.

μ–η^2:η^2

3.5 Models

The experimental and theoretical aspects of the entire span of spectral, magnetic and reactivity characteristics of many simple metal complexes have been thoroughly examined. These data have been invaluable for understanding in particular the metal site properties in metalloproteins. In many cases, it has been necessary to synthesize metal complexes of a new design or of an unusual oxidation state, in order to simulate that suspected for a protein metal site. Both mononuclear and polynuclear metal complexes have been prepared to compare with the increasingly complicated and unusual arrangements of metal ions being found in metalloproteins. There are numerous examples of the success of the model complex approach described throughout the book.

The power of the model approach in the structural characterization of metalloproteins is well illustrated by studies of binuclear copper (II) complexes relevant to oxyhemocyanin (Table 8.1). The size of the hemocyanin isolated from the lobster *Panulirus interruptus* (460 kDa, 6 subunits) presented considerable problems for crystallographers, although the structure has now been solved. Before this, comparison of the properties of Cu(II) complexes to those of oxyhemocyanin had been invaluable. It was known that in oxyhemocyanin:

(a) an O_2 molecule was bound symmetrically as a peroxide anion between the two copper (II) ions (resonance Raman: v (O–O) = 744–752 cm^{-1});

(b) oxyhemocyanin has absorption bands at 580 ($\varepsilon = 10^3$ M^{-1} cm^{-1}) and 340 nm ($\varepsilon = 2 \times 10^4$ M^{-1} cm^{-1}) which are the result of an $O_2^{2-} \rightarrow$ Cu(II) LMCT transition;

(c) the observed diamagnetism was due to strong antiferromagnetic coupling of the bridged copper ions ($-J < 300$ cm^{-1}), Cu–Cu distance = 3.5–3.7 Å (EXAFS).

Several peroxo dinuclear copper (II) complexes with different types of bridge linkages have been prepared and their properties compared to those of oxyhemocyanin. One exhibiting μ–η^2:η^2 type peroxo binding and for which a crystal structure was available had properties closest to those of oxyhemocyanin:

(a) v (O–O) = 741 cm^{-1};

(b) bands: 551 ($\varepsilon = 790$ M^{-1} cm^{-1}), 349 nm ($\varepsilon = 2.1 \times 10^4$ M^{-1} cm^{-1});

(c) diamagnetic and a Cu–Cu distance of 3.56 Å.

This complex is now known to have the same structure as the dicopper site found in oxyhemocyanin.

An intense effort has been made to prepare model complexes which mimic the functions as well as the spectral properties and metal site structure of the corresponding metalloprotein. Although there has been some success, the simple models rarely closely replicate the functionality and specificity of the natural systems. The contribution of the protein component cannot be underestimated! A new approach is the synthesis of models incorporating segments of protein. These more closely approach the structures of complicated protein systems and may more nearly act as functioning models. They have been likened to maquettes, which are small models used by sculptors and architects.

4 Structures and functions

4.1 Introduction

As we have seen, the spectroscopic properties of metalloproteins are vital for their monitoring and characterization as well as for illuminating the details of the metal site. Other properties are more directly related to the function of the metalloprotein. We can apportion these properties between those originating from the polypeptide and those arising from the metal ion. This is an artificial division for convenience and it should be emphasized that the two components will also often act synergistically resulting in a unique entity.

4.2 The role of protein

(a) The protein imposes a rigid and often unusual arrangement of ligands around the metal ion.

This can have mechanistic implications. In plastocyanin, there is an irregular but similar ligand geometry around the copper (I) and copper (II) forms (Section 3.2). This aids in electron transfer between the two oxidation states (in photosynthesis) because major structural changes in interconverting Cu(I) and Cu(II), which would slow down the rate, are unnecessary (Section 6.4). The importance of the protein in establishing this structural feature is confirmed by the observation of little structural difference between the native and apoproteins. In other words, the protein has done the trick!

(b) The protein provides a pathway for the protected or guided entry of substrates to the metal site.

In some cases, it is apparent from the crystal structure that there is, for the reactant, an easily accessible cleft leading to the metal ion situated at its base. In carbonic anhydrase, for example, the zinc ion is located at the base of a deep pocket, about 15 Å from the surface. In other metalloproteins access for a substrate may require (transient) thermal or allosteric broadening of a channel. The structures of nitrogenase and of the heme proteins myoglobin and cytochrome P-450 indicate no obvious channel for substrate entry. Indeed in P-450, the heme is buried deep inside the protein. In the case of a charged substrate, which is somewhat unusual, electrostatic guidance by certain amino acid side chains may be provided (Fig. 4.1). Only one side of the copper in bovine CuZn superoxide dismutase, SOD, (Section 6.3) is accessible to small molecules, which can pass through a conical channel that is narrower at the copper site than at the surface. Lining the channel near the surface are positively charged lys and negatively charged glu residues approximately 13 Å from the copper and they are believed to direct O_2^- ions into the channel. There is a network of residues in the channel which are

A type 1 copper centre is present in plastocyanin. This gives rise to an intense blue colour in the Cu(II) form. The crystal structure of poplar plastocyanin shows a distorted tetrahedral arrangement of donor atoms around the Cu from 2Ns (2 his) and 2Ss (cys, met). The colour arises from the cys → Cu(II) LMCT. Type 1 Cu proteins have a characteristic rhombic EPR signal with separate g_x, g_y and g_z values.

Q. There is a net negative charge on SOD at physiological pH. The rate of reaction with a negative ion such as O_2^{2-} should increase with increasing ionic strength according to kinetic theories. The rate is observed to decrease however. Why might this be?

involved in electrostatic guidance of O_2^- to the copper. Sucessive glu, lys, glu and thr sidechains are linked by hydrogen bonds (dashed line in Fig. 4.1). A positively charged arginine near the catalytic site may be used for docking the O_2^- ion next to the Cu^{2+} ion. Ligands attached to the metal ions are omitted in Fig. 4.1.

Q. Why would only lys residues at the mouth of the channel in superoxide dismutase be unproductive?

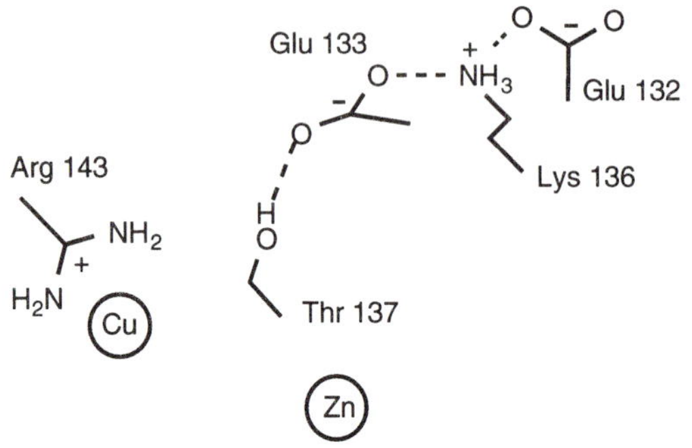

Fig. 4.1 Electrostatic guidance of superoxide ion to the Cu site in human superoxide dismutase. Two O_2^{2-} ions pass down the channel, react at the Cu site ($k\sim 10^9$ M^{-1} s^{-1}) and exit as O_2 and HO_2^{2-}. Rates were measured using pulse-radiolytic produced O_2^{2-} (Table 3.2).

(c) The main role of the protein, and one related to (a) above, is to provide the environment of the metal ion, specifically, the coordinated ligands and other nearby amino acids.

These could have a variety of duties. The ability of the protein to fine tune the properties of the metal centre is crucial to its function. This principle is demonstrated in the reaction centres of photosynthesis (Section 6.4) where the protein environment largely determines the redox potentials of pigment–protein complexes.

Hydrophobic clefts or patches in the protein can be used as a preturnover location of a hydrophobic molecule. Such a hydrophobic pocket, made up of valine, leucine and tryptophan residues, about 3–4 Å from the Zn in carbonic anhydrase has been suggested as the binding site for CO_2 prior to, or at,

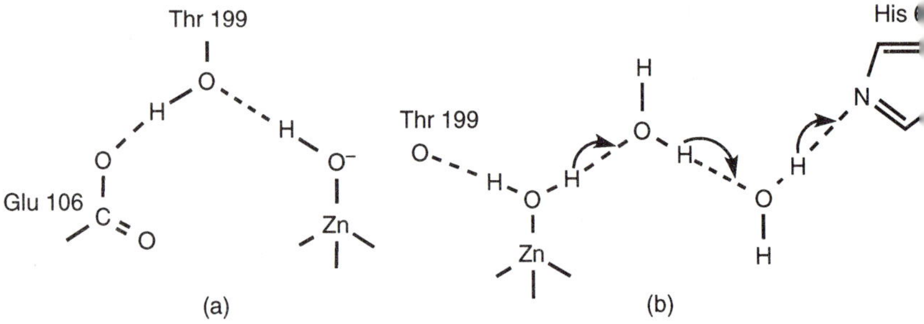

Fig. 4.2 Hydrogen-bonded solvent networks at the Zn site in carbonic anhydrase (see also Fig. 2.5). These play (a) orientation and (b) ionization roles in the catalysis by the metalloenzyme (Fig. 7.1). The arrows show the H^+ movements on ionization.

activation (Section 7.3). In addition, a hydrogen-bonded solvent network at the Zn site (Fig. 2.5) helps to locate the carbon dioxide molecule for reaction. The interactions shown in Fig. 4.2(a) optimize the geometry for $Zn–OH^-$ attack on CO_2 and (b) mediate the ionization of the $Zn–OH_2$ species (Section 7.3). Substitution of Zn by a number of metals (Co, Cu, Mn and Cd), and kinetic and NMR studies in addition to the crystal structure, have provided enough information to make carbonic anhydrase one of the best understood metalloenzymes (Section 7.3).

Examination of the crystal structure of yeast cytochrome c peroxidase, and of the stable O_2 adduct of an Fe(II) form of a mutant of the enzyme, as well as considerations of its biochemistry, show the influence of hydrogen bonding on the activation of O_2 by the enzyme (Fig. 4.3). An histidyl imidazole within the active site (his 52) aids in the transfer of a proton from the oxygen of H_2O_2 bound to iron (O_A) to the departing oxygen (O_B). The water leaves and hydrogen bonding occurs between the trp 51, arg 48 and the Fe=O entity. The behaviour of site specific mutants has been particularly useful in interpreting the roles of these residues.

Replacement of his 52 by a leucine residue decreases the rate of formation of compound I (eqn 6.6) by 5 orders of magnitude. Replacing arg 48 by lysine only has a twofold rate effect, suggesting that arg 48 stabilizes compound I but does not promote its formation.

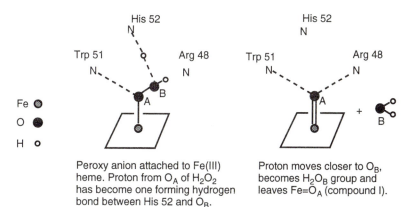

Peroxy anion attached to Fe(III) heme. Proton from O_A of H_2O_2 has become one forming hydrogen bond between His 52 and O_B.

Proton moves closer to O_B, becomes H_2O_B group and leaves Fe=O_A (compound I).

Fig. 4.3 Hydrogen bonding effects on the activation of O_2 by yeast cytochrome c peroxidase.

(d) Specialized molecules are required to transport and store metal ions in biological systems.

For the transition metals, only proteins are used for these tasks in vertebrates, but nonproteins are mainly used in lower organisms. For iron, the most well understood transition metal ion, transferrin and ferritin are used in animals for transport and storage, respectively, while in bacteria siderophores are employed for these purposes (Section 7.5). The s-block metal ions can be transported as aquated ions, but do require the intervention of ionophores, which are strong chelating ethers, or membrane proteins, for transfer across the central hydrophobic region of a membrane (Section 5.3).

4.3 The role of metal

A specific role for the metal ion can be assigned in a number of cases. These may include:

There are two fundamental types of coordination: (a) inner-sphere complexes in which coordinated water in the aqua ion $M(H_2O)_n{}^{m+}$ is replaced by the ligand and (b) outer-sphere complexes (or ion pairs) in which an undisturbed coordinated water molecule is between the metal ion and ligand, e.g.

$$M(H_2O)_6{}^{2+} + SO_4{}^{2-} \rightleftharpoons$$

$$M(H_2O)_6OSO_3 \rightleftharpoons$$

outer-sphere complex

$$M(H_2O)_5OSO_3 + H_2O$$

inner-sphere complex

(a) Enhancement of the stability of the biomolecule.

The s-block metal ions excel at this. Thus Mg^{2+} in particular, with its high intracellular concentration, is vital to the stabilization of DNA and RNA helices. The electrostatic repulsions between the negatively charged backbone phosphates in the polynucleotides are alleviated by coordination, albeit weakly by outer-sphere complexing, to Mg^{2+}. All enzyme catalysed reactions involving DNA and RNA polymerases require divalent cations, usually Mg^{2+}. Calcium ions stabilize some proteins, e.g. thermolysin (a peptidase), toward unfolding (denaturing) on heating. The other metal present (Zn^{2+}) in thermolysin has a catalytic role (see later). Tungsten-containing proteins are very rare and are found mainly in thermophilic anaerobes. *Pyrococcus furiosus* grows optimally at 100 °C. The W-containing enzyme, aldehyde ferredoxin oxidoreductase isolated from it, is extremely thermally stable. It was a disappointment to the authors to learn that the stability of such metalloenzymes is due to a number of factors (salt links, H bonding, low surface area to volume ratio) and not mainly to the presence of the metal ion!

(b) Promotion of an essential conformational change in the protein.

There are many calcium binding proteins with negatively charged regions because they are rich in glu and asp residues. In some the binding of Ca^{2+} ion causes a conformational change in the protein allowing it to interact with another protein and thereby trigger enzyme activity; e.g. calmodulin will activate NO synthase, phosphorylase kinase and (Ca^{2+},Mg^{2+}) ATPases (Fig. 5.4).

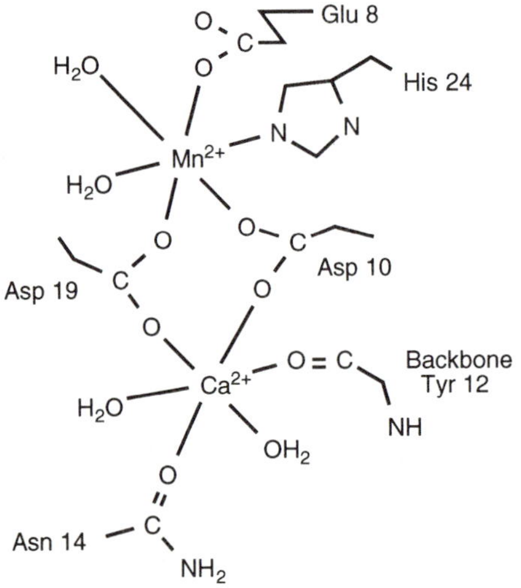

Fig. 4.4 The Ca^{2+}, Mn^{2+} site in jack bean concanavalin A. Each metal ion exhibits nearly octahedral coordination and includes two waters which are hydrogen bonded to nearby residues.

Binding of Ca^{2+} or Mg^{2+} to M^{2+} sites in troponin C stabilizes the structure as in (a) above, but may also cause localized structural changes which trigger muscle contraction (Section 5.5). In concanavalin A (con A), a manganese ion helps to position the ligands correctly around the calcium ion (Fig. 4.4). This in turn ensures the correct orientation of amino acid residues involved in binding the centre of a methyl α-D-pyranoside molecule 8.7 Å from the Ca^{2+} (and 12.8 Å from the Mn^{2+}, Fig. 4.5). In the con A–sugar complex, the sugar is in the chair conformation, held there by extensive hydrogen bonds and van der Waals forces (Section 2.4), some of which involve Ca^{2+} ligands (tyr 12, asn 14, arg 228). In a final example, in transcription factor III A, zinc ion binds tetrahedrally to 2 cys and 2 his residues and, in so doing, promotes formation of a loop or 'finger' (Fig. 4.6) which fits into the major groove of DNA. We shall have more to say about these most important transcription factors later (Section 7.6).

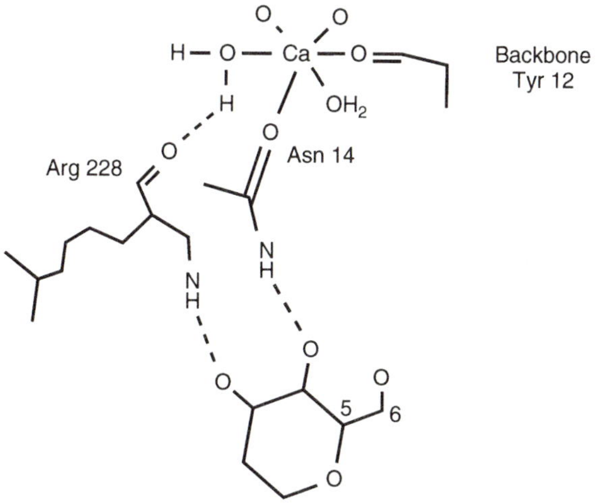

Fig. 4.5 Methyl-α-D-mannopyranoside binding site in jack bean concanavalin A. Only the amino acids which are coordinated to the Ca^{2+} and associated with the sugar by hydrogen bonding (asn 14 and arg 228) or by hydrophobic interactions (tyr 12 with C5 and C6) are shown.

There are 100s of lectins, which are sugar-binding proteins. Those from legumes, such as conA from the seeds of jack beans, contain Ca^{2+} and Mn^{2+}. Their function in plants is unknown. Lectins from animals contain only Ca^{2+} (C-type) which is coordinated to the sugar. These lectins act as mediators of cell recognition, e.g. in the adhesion of viruses to host cells. Lectins are useful for carbohydrate purification and in probing membrane structure.

(c) Modification of the function of the biomolecule.

MerR is a metalloregulatory protein which switches on the transcription of the bacterial mercuric ion resistant genes when Hg^{2+} is present at as low as nM concentration. Mercury (II) ion binds specifically and tightly to the DNA-bound MerR causing a DNA conformational change. This facilitates the binding of RNA polymerase and thus converts MerR from a repressor to a strong activator of transcription.

(d) Metal ions or complexes can act as cofactors.

Many enzymes require the presence of a +1 cation, particularly K^+, for activity. Such ions probably bind to the enzyme some distance from the substrate site and act by stabilizing a particular conformation. Potassium ion is the most effective of the alkali ions for promoting catalysis by pyruvate kinase (see Table 5.3). The cation is believed to link the substrate to the enzyme surface, thereby stabilizing the enzyme–pyruvate complex.

Nearly all the roles described in Sections 4.2 and 4.3 are used at some time or other to bring about the rate enhancements which are promoted by some metalloproteins. This leads us to one of the most important functions exhibited by metalloproteins, namely catalysis.

Xenopus laevis is a species of frog often used in research because of the size and robustness of its eggs. Transcription factor IIIA is required to activate the gene that expresses 5S RNA. It binds both to specific DNA sequences within the 5S RNA gene and also to the 5S RNA molecule. It was the first member of the zinc finger family to be discovered.

The sequence: cys*-x$_{2-4}$-cys*-x$_3$-phe-x$_5$-leu-x$_2$-his*-x$_{3-5}$-his* occurs in each of the nine units and appears in at least 200 different proteins. The ligands marked * and phe and leu (hydrophobic residues which stabilize the fingers) are invariant.

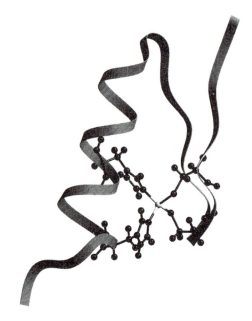

Fig. 4.6 A zinc finger in transcription factor IIIA from *Xenopus laevis*. This is one of nine similar units in tandem. The right hand folds back on itself to form a two-strand structure (β-sheet); the left hand twists into an α-helix. Two cys residues in the β-sheet and two his in the α-helix bind tetrahedrally to the Zn.

4.4 Metalloenzymes

Monoclonal antibodies specific for a distorted N-methyl porphyrin catalyse the metal ion incorporation into the corresponding planar non-methylated porphyrin, probably by forcing the latter into the distorted conformation which is known to chelate more rapidly. The catalytic properties of this abzyme resemble those of ferrochelatase, which catalyses the biological insertion of Fe(II) into protoporphyrin IX, also probably by ring distortion.

Enzyme classification (EC): each enzyme is assigned a set of numbers which reflect its main class, details of the reaction catalysed and its serial number. Carbonic anhydrase, for example is EC 4.2.1.1.

The catalysts of Nature are the enzymes. Metalloenzymes have been estimated to make up about 40% of all enzymes and are represented in each of the internationally recognized six classes. Table 4.1 shows the six basic types of reactions catalysed. Specific examples of these are presented in Tables 5.3, 6.3, 6.4, 7.1 and 7.2. Metalloenzymes range in size from the very small (only about 80 or so amino acids in cytochrome c_{551}) to very large entities such as glutamine synthetase, MW = 619 kDa, consisting of 12 subunits arranged at the corners of a hexagonal prism and which has two M^{2+} (Mg^{2+} *in vivo*) ions in each subunit. Uncatalysed reactions are much slower and their rates vary more widely than the catalysed ones. Some enzymes are able to enhance the rate of the spontaneous reaction by a factor as large as 10^{17} (Fig. 4.7). Antibodies, which are large proteins, can also catalyse a wide variety of chemical reactions. Abzymes, as they are termed, are not usually as effective as enzymes, but can produce 10^3–10^6 fold enhanced rates over the uncatalysed reaction. The metal ions in metalloenzymes can participate in the catalysis in several ways, which are usually coupled to, and reinforced by, nearby protein structural effects. The metal ion can:

(a) Act as a collecting point for the reactants.

Table 4.1   Classes of enzymes

Class of enzyme	Reactions catalysed
1. Oxidoreductases	Wide variety of oxidations and reductions
2. Transferases	Transfer of functional group from one substrate to another, e.g. kinases, which aid phosphoryl group transfer between ADP and ATP
3. Hydrolases	Hydrolysis, e.g. peptide bonds (peptidases and proteases); phosphate esters (phosphatases)
4. Lyases	Addition or removal of a group, e.g. H_2O (hydratases) and enolases
5. Isomerases	Interconversion of isomers, e.g. optical forms (racemases)
6. Ligases	Bond formation coupled to breakdown of ATP

A number of enzymes require a non-protein entity to be functional. In such cases the enzyme is referred to as a holoenzyme and consists of apoenzyme plus a non-protein constituent. Various terms have been used for the latter including coenzyme, but also prosthetic group (usually if tightly bound) or cofactor, if it is a loosely bound organic or inorganic molecule. One or all of these terms have been applied to NAD, flavin nucleotides, coenzyme A, heme and even the metal ions in metalloproteins.

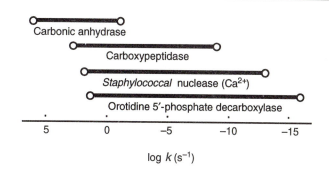

Fig. 4.7 First order rate constants at 25 °C for four spontaneous and catalysed hydrolysis or decarboxylation reactions.

Orotidine 5′-phosphate decarboxylase catalyses the last step in the biosynthesis of pyrimidines. The spontaneous rate is so very slow ($t_{1/2}$ estimated as > 8 million years at 25 °C or 10 years at 100 °C) that a primitive catalyst from a thermophylic organism must have been around in prehistoric times.

This function is analogous to the neighbouring group effect, which is well known in organic chemistry. Invariably, one reactant will be intimately associated with the metal ion, probably coordinated to it. Another reactant will be located near to, and with the correct orientation for attack at, the metal ion. A hydrophobic patch or hydrogen bonds, provided by neighbouring amino acids, may anchor one reactant close to the metal ion (Section 4.2). In an extreme case, the substrate may be radically altered. Thus, D-xylose is forced into an 'open-chain' conformation, rather than the usual cyclic form, when attached to D-xylose isomerase. Terminal hydroxyl groups on the sugar, away from the site of the isomerization, are bound to one of two manganese ions in the enzyme (Table 7.2). The other manganese ion has a direct role in the isomerization.

(b) Enhance the acidity of a coordinated ligand.

The positive charge on the metal ion will, for example, aid in the ionization of a coordinated water, i.e., lower the pK_a compared to that of bulk water. This ensures a relatively high concentration of (coordinated) hydroxide ion. Although this is a poorer nucleophile than free OH^-, it is a much more effective nucleophile than water and is thus available for accelerated nucleophilic reactions at *neutral, physiological* pH where the concentration of OH^- is only ~0.1 μM. Another example is in the enhancement of the C–H acidity in mandelate, promoting the formation of an enol intermediate and thus (*R*)-, (*S*)-interconversion (Table 5.3).

(c) Serve as an oxidation–reduction centre for the manipulation of electrons in the oxidoreductases (Section 6.3).

The metal ion can also bind an oxidant such as O_2 and modify its electronic configuration so as to enhance oxidative reactivity. In the heme proteins, the entire metal–porphyrin unit plus the axial group or groups play an important role in the function of the enzyme (Section 6.3).

(d) Play a structural role (see Section 4.3 (a)).

Bivalent Zn^{2+}, being a d^{10} species with completely filled d-orbitals, has some of the properties of alkaline earth bivalent cations and therefore can have, like them, a structural as well as a catalytic role in metalloproteins. This is well illustrated by liver alcohol dehydrogenase which catalyses the reaction:

$$C_2H_5OH \ + \ NAD^+ \ \longrightarrow \ CH_3CHO \ + \ NADH \ + \ H^+ \qquad (4.1)$$

and removes alcohol both produced internally and consumed externally. Each subunit of the dimer binds one NAD^+ and two very different Zn^{2+} ions. One Zn^{2+} is bound to four cys residues, is inaccessible to solvent and substrate and undoubtedly has a structural function. The other Zn^{2+} is bound tetrahedrally to two cys, his and H_2O (or OH^-) in a deep pocket (Section 4.2(b)) about 20 Å from the surface. It affects catalysis in ways outlined above. It attaches and lowers the pK_a of the alcoholic OH (Section 4.4(b)) facilitating hydride transfer from the alcohol to the nicotinamide ring of the NAD^+ coenzyme (Fig. 6.2). It also helps to position the substrate for reaction (Section 4.4(a)).

Some microorganisms thrive in extreme conditions (extremophiles). Enzymes derived from them, termed extremozymes, must therefore function at high temperature and pressure (deep-sea hydrothermal vents) or near or below 0 °C (arctic waters), in high salt concentrations (e.g. 3–5 M NaCl in the Dead Sea) low pH (geothermal springs) or high pH (sewage sludge). Extremozymes studied so far have similar activities and structures to their mesophilic analogues isolated in moderate conditions. Subtle, and as yet incompletely understood differences must account for their striking variation in stability. The various enzyme classes (Table 4.1) are well represented and a number of extremozymes contain metal ions.

5 The s-block

5.1 Introduction

In the elements of groups one and two, there are one s- and two s-outer electrons respectively, which can be easily lost to give the predominant +1 and +2 oxidation states associated with these groups. Ionic bonds therefore predominate in compounds in which these elements are present. The radii of the ions of biological relevance are shown in Table 5.1.

Table 5.1 Ionic radii of biologically important cations of the s-block

Ion	Radius (Å)	Ion	Radius (Å)
Na^+	1.02	Mg^{2+}	0.72
K^+	1.38	Ca^{2+}	1.00

The values refer to the 'bare' ion denuded of any solvation. The aquated ion includes inner shell coordination and outer shell attraction of solvent molecules. As these ions have no unpaired electrons, they are diamagnetic and colourless and therefore monitoring them presents formidable problems. The use of NMR (^{23}Na) or replacing the metal with more easily analysed ones (Ca^{2+} by Eu^{3+}) have been profitable techniques. However, for the detection of rapid changes in concentration (important for the analysis of Ca signalling), the use of fluorescent indicators that specifically bind to Ca^{2+} and thereby attenuate the observed fluorescence, has been especially helpful. The s-cations are generally poor complexers (Ca^{2+} is by far the best) and require strong binding, and therefore, chelating or macrocyclic ligands or proteins to form very stable entities. They are hard cations (Section 2.3) and generally have an affinity for O donor ligands, although Mg^{2+} is coordinated to N donors in some very important molecules. The aquated Na^+, K^+ and Ca^{2+} ions are extremely labile, that is their coordinated waters exchange rapidly with ligands in aqueous solution, $k \sim 10^8\,s^{-1}$. By contrast, for Mg^{2+}, $k \sim 10^5\,s^{-1}$, a decrease which has biological implications.

Calcium ion indicators must function at neutral pH and be sensitive to small Ca^{2+} concentration changes, such as 0.1 μM (resting cell), increasing to 1–10 μM after a message is transmitted. Certain fluorescent dyes or photoproteins like aequorin (from jellyfish) which emit light on binding Ca^{2+} are responsive even in the presence of large concentrations of Mg^{2+}, Na^+ and K^+. Using a complex fluorescence microscope and living cells or tissue on the microscope stage, it is possible to follow the fluctuations of Ca^{2+} concentrations *in vivo* (heartbeat, external stimulations).

5.2 Biological aspects

The group 1 and 2 cations, in addition to Cl^- and HPO_4^{2-} anions, provide the background electrolytes for living systems, Table 5.2. Since they can exist as aquated ions, M^+_{aq} and M^{2+}_{aq} at physiological pHs, they can act as charge carriers and promote charge gradients across membranes. Magnesium and especially calcium have a much richer biochemistry than sodium or

Table 5.2 Approximate mammalian cellular concentrations

Ion	Intracellular (mM)	Extracellular (mM)
Na^+	10	150
K^+	100	5
Mg^{2+}	2.5	1.5
Ca^{2+}	0.1^a	2.5
Cl^-	50	100

a In the cytoplasm of the resting cell, the concentration is only ~0.1μM.

Q. How can the larger concentrations of K^+ inside and Cl^- outside the cell be rationalized? (Hint: consider proteins and polynucleic acids trapped within the cell).

Cells are working components of tissues and organs. There are two main types of cells. Prokaryotic cells have no membrane-bound nucleus or organelles. They are present only in bacteria. *Escherichia coli* lives in human gut and is widely used as a representative prokaryote. All other organisms (animals, plants, fungi and algae) contain eukaryotic cells which are larger and more complex (Fig. 6.1).

potassium, and act mainly intracellularly in eukaryotes. Their role in prokaryotes is still uncertain. With the exception of Na^+, the s-block cations tend to function mainly as stabilizers of biomaterials (e.g., Mg^{2+} with nucleic acids, Section 5.4) and stabilizers of, and initiators of conformational changes in, proteins (Ca^{2+} with numerous proteins). Calcium helps in the unfolding of trypsinogen and the subsequent cleavage to trypsin, which is then stabilized toward hydrolysis by Ca^{2+} binding. Magnesium and calcium ions are used to crosslink and stabilize the outer cell wall, accounting for the ability of chelating agents to disrupt cell membranes by reacting with and thus removing these metal ions. Cations of the s-block, except Ca^{2+} (Section 5.5), are rarely found as the tightly bound metal in a metalloenzyme and in this respect they differ markedly from the d-block cations (Chapters 6 and 7). However, they do bind weakly to proteins and thereby play an essential role, either by linking substrate (S) to the enzyme (E) [S–M^{n+}–E] to help orientate it for reaction or by modifying the substrate in forming the entity [M^{n+}–S–E]. Although intracellular roles for calcium predominate, the higher concentration of extracellular Ca^{2+} (Table 5.2) allows it to associate with weakly binding proteins such as prothrombin, which can bind to certain cell membrane platelets in the presence of Ca^{2+}. This is an essential step in the complex chain of interconnected processes, involving Ca^{2+} and many enzymes and proenzymes, which leads to blood coagulation.

5.3 The transport of ions

It is still something of a mystery that H_2O (certainly hydrophilic!) passes very rapidly through the hydrophobic membrane.

Metal ions must frequently be moved between different parts of multicellular organisms. In the body, alkali metal ions in particular, but also alkaline earth cations, can be transported in serum as the aquated ion because they are soluble at physiological pH and therefore do not precipitate as the aquated transition metal ions would. The different distribution of cations inside and outside the cell (Table 5.2) means that metal ions must therefore be constantly transferred across the plasma membrane (Fig. 5.1) which encloses the cell, to maintain homeostasis and a healthy cell. Transport is no problem for gases (e.g. O_2 and N_2) and organic compounds like steroids since these are membrane soluble. Small molecules such as H_2O, urea and anions (e.g. Cl^-, HPO_4^{3-}) can also move through transient gaps in the membrane, or through tunnels in the membrane proteins, by diffusion from regions of higher to lower concentrations. The transfer of the hydrophilic cations Na^+, K^+, Mg^{2+} and Ca^{2+} across the central hydrophobic region of the membrane is quite another matter and is still incompletely understood. There are in essence three transmembrane transport mechanisms for these cations:

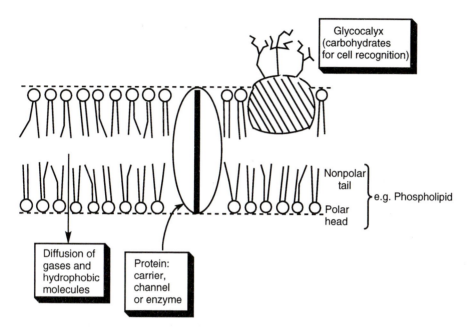

Fig. 5.1 Edge of a plasma membrane. Lipid bilayers are dynamic and fluid since they are composed of many separate molecules. The membrane separates the protoplasm from a hostile environment and in addition regulates the transport of species in and out of the cell.

(a) The cation may be encapsulated within a ligand, called an ionophore (Fig. 5.2), resulting in the formation of a complex with many organic groups (e.g. –Me) on its periphery. These complexes, because of their hydrophobic sheath, can diffuse across the fatty-membrane, although they cannot function at temperatures below which the membrane becomes stiff. This problem does not arise with the next mode of ion transfer.

(b) A channel or tunnel may exist in a polypeptide or protein which spans the membrane. The cation will move down the hydrophilic channel, perhaps with transient interactions with donor atoms (carbonyl groups) from the polypeptide. Gramicidin forms such a channel. It occurs in a variety of conformations depending on the physical environment; i.e. solid or in organic solvents. In the phospholipid membrane, a channel is made up of dimers of single stranded β-helical gramicidin monomers in which terminal formyl NH groups of each monomer are H bonded. It forms a central hole 4 Å in diameter and about 50 Å in length spanning the lipid bilayer and allowing the passage of *monovalent* (including H^+) cations (Fig. 5.3).

These methods, (a) and (b), are 'passive' since ions flow in the direction of the concentration gradient. In contrast, the third method of transfer, (c), termed active ion pump, moves ions against the normal distribution and therefore requires the input of energy from the cell.

Ionophores are obtained from fungi and marine organisms. They act as antibiotics by opening a channel to ion transfer, thereby upsetting ion balance and killing the cell. Since they cannot distinguish microbial from mammalian membranes, they are therapeutically useless! Some charged ionophores have however become very useful in veterinary medicine. The cells of *coccidia* , an internal poultry parasite, when treated with monensin literally blow up. This results from the ionophore allowing loads of Na^+ to move into the cell (and replace H^+).

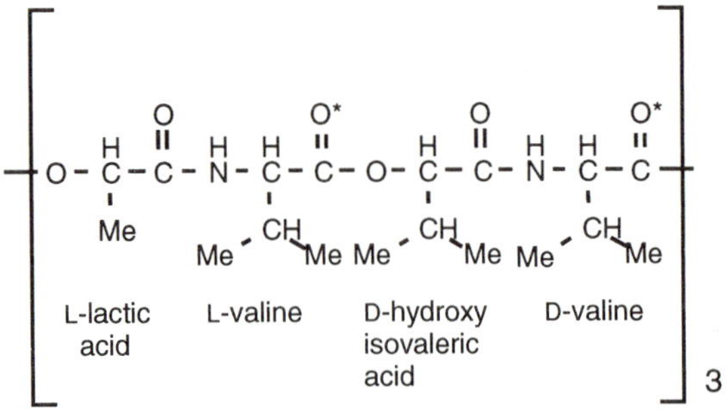

Fig. 5.2 Valinomycin. A product of *streptomyces* fermentation and the first ionophore isolated. It is a cyclic depsipeptide, containing both ester and amide bonds. Selective, it is an almost perfect fit for K^+, which is surrounded octahedrally by six O*.

Gramicidin channels are useful as models for the examination of lipid–protein interactions, the mechanisms of ion permeation and conformational changes in ion-permeable channels.

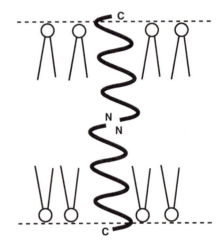

Fig. 5.3 Gramicidin A structure in a phospholipid membrane. Gramicidin A is a 15 residue polypeptide with alternating L and D residues produced by *Bacillus brevis*.

Ion pumps (ATPases) are large, complex membrane systems. Their actions are interrelated. Any inhibitor of Na^+/K^+ ATPase will retard the exit of Na^+ from the cell. This means that the Na^+/Ca^{2+} interchange will be stimulated (via a Na^+/Ca^{2+} ATPase pump), leading to enhanced Ca^{2+} entry into the cell. One such inhibitor is digitalis, from purple foxglove. By increasing the Ca^{2+} concentration, digitalis can regulate and strengthen a heartbeat and has long been used as a cardiac drug.

(c) Here a membrane-bound enzyme is able to 'pump' ions across a membrane between *intra*cellular and *extra*cellular space. For example, Na^+/K^+ ATPase catalyses the transfer:

$$3Na^+ \text{ (intra)} + 2K^+ \text{ (extra)} + Mg^{2+}ATP^{4-} + H_2O \rightarrow$$
$$3Na^+ \text{ (extra)} + 2K^+ \text{ (intra)} + Mg^{2+}ADP^{3-} + HPO_4^{2-} + H^+ \quad (5.1)$$

(Note the involvement of Mg^{2+} ions with the nucleotides). The net flow of K^+ into and Na^+ out of the cell is counter to the normal concentration gradients, but constantly required. The necessary energy is derived from Mg^{2+} catalysed ATP hydrolysis (Section 8.10). There are a number of different ion pumps at various locations, coupled so as to give movement in the same (symport) or opposite directions (antiport). The Na^+/H^+ antiport uses respiratory H^+ gradients (Section 6.4) to generate transmembrane gradients and acts to maintain pH. An H^+/K^+ antiport coupled to a K^+/Cl^- symport is used to maintain stomach acidity (pH $\approx$ 1). Once inside the cell, these cations require ligands or proteins to carry them for use or storage. Since Ca^{2+} serves as a vital secondary messenger (Section 5.5) the initiation, propagation and termination of the calcium signal is particularly complex. Several transport proteins (calcium binding proteins), storage proteins (calsequestrin) and membrane translocation proteins (Ca^{2+}ATPases) are required and the processes involved are far from understood.

5.4 Magnesium, nucleotides and nucleic acids

One of the major functions of Mg^{2+} arises from it being able to interact with anions in nucleotides (ATP, ADP) and polynucleotides (nucleic acids, DNA and RNA). Indeed most (90%) of intracellular Mg^{2+} is bound to ribosomes (the RNA-protein complexes that mediate protein synthesis) or polynucleotides. In the absence of Mg^{2+} ion, many RNAs exist in partially denatured states. Addition of Mg^{2+} initiates formation of the native structure within minutes. Both DNA and RNA polymerases, which catalyse, respectively, the replication and transcription of DNA, have a specific requirement for Zn^{2+} as well as another divalent cation, usually Mg^{2+} but sometimes Mn^{2+}. One can in fact study some Mg^{2+}–containing proteins by replacing Mg^{2+} with the EPR active Mn^{2+} ion. The association of Mg^{2+} with nucleic acids is necessary for nerve impulse transmissions, the metabolism of carbohydrates and muscle contraction. It was only about ten years ago that it was shown that a non-protein, namely RNA, could function as an enzyme. In all ribozymes (as they are called) so far examined, a divalent metal ion, often Mg^{2+}, is essential, so that these represent a new class of metalloenzyme. They catalyse a variety of reactions involving RNA and DNA molecules. Mg^{2+} features in the chemistry in ways similar to those seen in metalloproteins, as well as helping the secondary structure of RNA to fold into its active tertiary structure. The metal ion stabilizes the structure by neutralizing the charges on the extended polyanions. The study of the folding pathways in proceeding from a linear polyanionic chain of nucleotides to the final compact structure of a ribozyme induced by adding Mg^{2+} ion, showed that the mechanism resembled that of multidomain proteins. That is, formation of independent stable substructures preceeds the folding into their final conformation. The resolution of a recent crystal structure of the hammerhead ribozyme substrate complex was too low to detect the Mg^{2+} ion, so its role is still uncertain. However, catalytic (Lewis acid cleavage of a P–O bond) and stabilizing functions have been proposed.

Bacterial exonuclease III is used in DNA repair. The enzyme searches for damaged sites along the DNA helix. It then 'flips out' the incorrect bases (misincorporated during DNA synthesis or chemically damaged) from the interior of the DNA helix and cleaves the backbone. The cut DNA is then corrected and respliced by other enzymes. Exonuclease III binds to the flipped out base in a pocket near to the catalytic Mg^{2+}.

The entry of excess Ca^{2+} into heart or arterial smooth muscle can promote excessive excitation. Channel blockers can inhibit this entry and act as calcium antagonists. Nifedipine (a 1,4-dihydropyridine derivative) is used in the treatment of angina pectoris and high blood pressure.

Mg^{2+} plays a significant role in the binding of small molecules to DNA. The antibacterial quinoline, Norfloxacin, both intercalates and binds to the DNA surface. Two such molecules are held together by two Mg^{2+} bridges. Each Mg^{2+} ion coordinates to the carbonyl and carboxyl Os of the drug and with one oxygen on DNA phosphate as well as two H_2O ligands. Such observations are likely to provide a handle on the mechanism of action of the drug and to lead to more effective drugs.

It is thus clear that Mg^{2+} ion has important catalytic (in which it acts as an innocuous Lewis acid) and structural roles in enzymes which catalyse a variety of reactions. A selection of enzymes and the reactions they catalyse are shown in Table 5.3.

Table 5.3 Varied roles of magnesium in enzymes

- *Enolase:* (yeast) Five-coordinate Mg^{2+} in the enzyme aids removal of OH^- by complexing substrate:

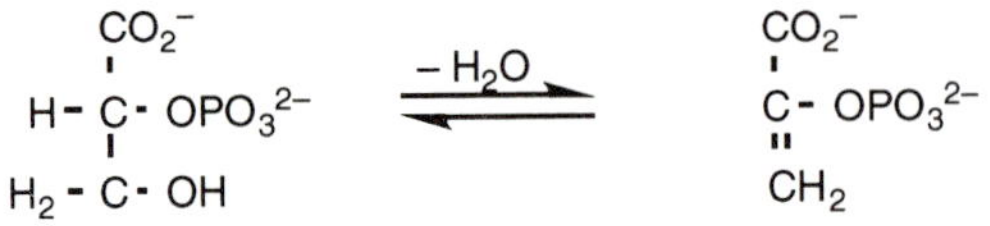

$$\begin{array}{c}
CO_2^- \\
| \\
H-C-OPO_3^{2-} \\
| \\
H_2-C-OH
\end{array}
\quad \xrightarrow{-H_2O} \quad
\begin{array}{c}
CO_2^- \\
| \\
C-OPO_3^{2-} \\
\| \\
CH_2
\end{array}$$

Part of a chain linking glucose to pyruvate (see next enzyme)

- *Pyruvate kinase:* uses K^+ for orientating substrate and two Mg^{2+} to chelate and help position ADP relative to substrate. All kinases require M^{2+}, which is usually Mg^{2+}.

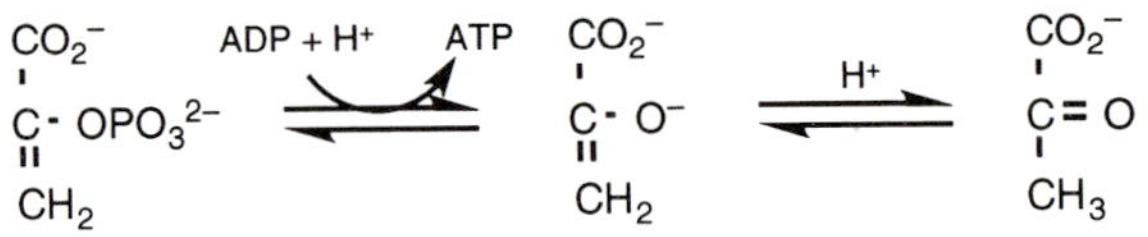

$$\begin{array}{c}
CO_2^- \\
| \\
C-OPO_3^{2-} \\
\| \\
CH_2
\end{array}
\quad \underset{ATP}{\overset{ADP+H^+}{\rightleftharpoons}} \quad
\begin{array}{c}
CO_2^- \\
| \\
C-O^- \\
\| \\
CH_2
\end{array}
\quad \xrightarrow{H^+} \quad
\begin{array}{c}
CO_2^- \\
| \\
C=O \\
| \\
CH_3
\end{array}$$

It is the final reaction of the chain linking glucose to pyruvate.

- *Mandelate racemase:* (*Pseudomonas putida*) Mg^{2+} is coordinated to H_2O, three amino acid residues and (when substrate is bound) the hydroxyl and carboxylate oxygen of mandelate. Proton removal and addition are aided by his and lys. Enolic intermediate is stabilized by interaction with electrophilic groups in the active site of the enzyme (glu, lys and Mg^{2+}).

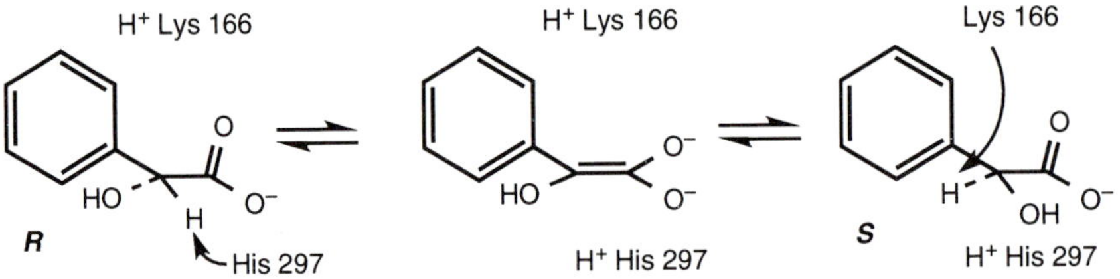

- *Ribulose 1,5-biphosphate carboxylase:* (tobacco) Mg^{2+} known to be five coordinate from the crystal structure. The Mg^{2+} probably stabilizes the enediolate intermediate.

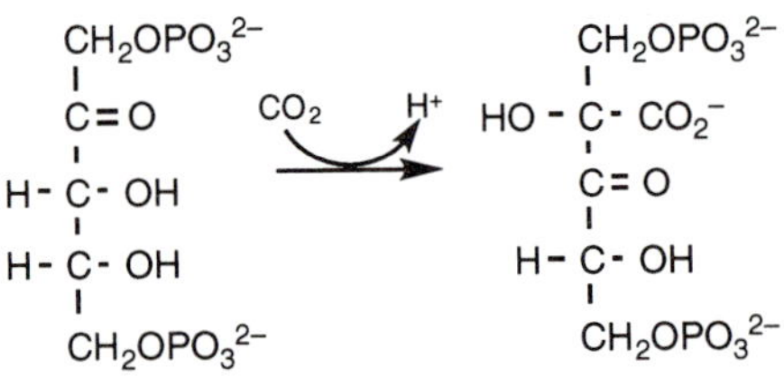

$$\begin{array}{c}
CH_2OPO_3^{2-} \\
| \\
C=O \\
| \\
H-C-OH \\
| \\
H-C-OH \\
| \\
CH_2OPO_3^{2-}
\end{array}
\quad \xrightarrow[H^+]{CO_2} \quad
\begin{array}{c}
CH_2OPO_3^{2-} \\
| \\
HO-C-CO_2^- \\
| \\
C=O \\
| \\
H-C-OH \\
| \\
CH_2OPO_3^{2-}
\end{array}$$

Involved in CO_2 fixation in Calvin cycle, the most abundant protein in the biosphere.

5.5 The versatility of calcium

Calcium as a messenger

A primary messenger is an extracellular agent, e.g. a hormonal or electrical signal, which initiates a response to an extracellular occurence. The signal is transmitted via a secondary messenger. Calcium acts as a secondary messenger intracellularly because its concentration can change rapidly in response to external stimuli. This concentration change is controlled by a Ca^{2+} binding protein. One such protein is troponin C, which is part of the troponin complex of skeletal muscle thin filaments. A change in concentration of free Ca^{2+} from 10^{-7} M (resting cell) to 10^{-5} M (stimulated by calcium entry) will induce binding of calcium to troponin C. This binding causes a conformational change in the protein which is transmitted to other protein components of the thin filaments, ultimately resulting in muscle contraction. The requirement of calcium for muscle contraction has been known for over a century. One other Ca^{2+} binding protein of note is calmodulin, whose function, like that of troponin C, is triggered by a change of calcium ion concentration which initiates the activation or deactivation of a large number of proteins and enzymes (Fig. 5.4). Examples of such enzymes are NO synthase (Section 8.9), NAD kinase used in the synthesis of NADP and adenylate and guanylate cyclases which catalyse the formation of

Troponin C, 18 kDa, contains two high affinity Ca^{2+} sites ($K > 10^6$ M^{-1}) that are always occupied by two Ca^{2+} ions which serve structural purposes. Two low affinity sites are the locations of the calcium ion stimulation. There are now more than 200 known calcium binding proteins of which 30 or more have had crystal structural determinations. Many share common structural motifs (calmodulin fold).

Q. What value for K_s (stability constants) can account for free Ca^{2+} at 0.1 μM and bound Ca^{2+} at 1–10 μM?

Q. How does Ca accomplish its biological functions in an environment of Mg^{2+} which is 10^3 times higher in concentration?

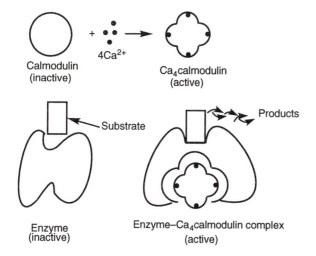

Calmodulin (inactive) $+ 4Ca^{2+} \rightarrow$ Ca$_4$calmodulin (active)

Substrate → Enzyme (inactive)

Products ← Enzyme–Ca$_4$calmodulin complex (active)

Fig. 5.4 Triggering of enzyme activity by Ca^{2+} ions binding to calmodulin. The calmodulins are small proteins with glu and asp residues which bind Ca^{2+} ions. The protein conformation is changed on metal ion binding so that it can now reorganize, bind and activate an enzyme. The details are still not understood and this is hampered by the lack of a structure of calmodulin without bound Ca^{2+} ion.

A single amino acid (glu 41) drags a portion of troponin C toward a newly bound Ca^{2+} at a low affinity site and promotes a conformational change. This is demonstrated by the fact that the E41A mutant is ineffective.

cAMP and cGMP (Section 8.10). The versatility of calcium in inducing a variety of intracellular processes probably arises because there are many ways in which a polypeptide can fold around the metal ion. The calcium binding proteins must associate specifically and rapidly with Ca^{2+} since K^+ and Mg^{2+} ions are present in much higher concentrations (Table 5.2).

The digestion of starch begins with saliva in the mouth and continues in the small intestine under the influence of α-amylases, which convert it to small glucose polymers. Final conversion to glucose requires other enzymes. α-Amylases are used in paper manufacture and beer production.

Calcium in extracellular enzymes

Calcium ion can also form an integral part of a number of extracellular enzymes. Since the extracellular concentration of Ca^{2+} is millimolar (Table 5.2) it will bind substantially to a biomolecule even if the affinity constant is as low as $10^4 M^{-1}$. The metal ions may help stabilize the molecule as we have observed with trypsin (Section 5.2). The three Ca^{2+} ions present in one of the three domains formed by the folding of the barley α-amylase polypeptide chain are vital to the maintenance of the structural integrity of that domain. One Ca^{2+} ion is close to the surface of the protein and may have a catalytic role. Phospholipases are a group of digestive enzymes which catalyse the hydrolysis of membrane phospholipids. They are present in high concentrations in intestinal juices and venoms. Catalysis is undoubtedly a function of Ca^{2+} in pancreatic phospholipase A_2. This enzyme contains an hydrophobic channel that allows a phospholipid aggregate to reach the active site of the enzyme. Here there are key amino acids and a Ca^{2+} ion, which is believed to stabilize an oxyanion transition state. In these enzymes, Ca^{2+} shows its versatility with coordination numbers (O donor ligands) of 6, 7 and 8.

Calcium biominerals and biomineralization

Calcium from the blood stream can precipitate as a crust on damaged blood vessels and in the heart. This restricts blood flow, raises the blood pressure and can lead to heart attacks. The chelating ligand EDTA (Section 7.8) in the form of $Na_2MgEDTA$ has been suggested for use in dissolving these deposits, thus obviating expensive bypass surgery which is the usual method for correcting this serious problem.

Calcium is a major component of bones, teeth and shells largely because of the insolubility of calcium carbonates and phosphates at physiological pH (Table 5.4). Biominerals can be produced both internally and externally by living organisms and may be crystalline or amorphous. An inorganic/organic

Table 5.4 Inorganic solids of group 2 in biological systems

Chemical	Mineral	Occurence
$MgCO_3$	Magnesite	Coral skeletons
$CaCO_3$	Aragonite	Shells (and pearls) of molluscs
	Calcite	Bird egg shells; gravity device in inner ear
$CaCO_3 \cdot nH_2O$	Amorphous	Ca storage in plants
$Ca(C_2O_4) \cdot nH_2O$	Whewellite ($n = 1$)	Ca storage in plants; stones in kidney or
	Weddelite ($n = 2$)	urinary tract
$Ca_{10}(OH)_2(PO_4)_6$	Hydroxyapatite	Bones and teeth in vertebrates
$CaSO_4 \cdot 2H_2O$	Gypsum	Gravity device in jellyfish
$SrSO_4$	Celestite	Exoskeletons of certain plankton
$BaSO_4$	Baryte	Gravity device in algae

Human saliva contains a supersaturated basic calcium phosphate solution which is used for recalcification and protection of tooth enamel. Precipitation is prevented by a Ca-binding protein, statherin.

composite invariably results from the process, and this combination leads to the desired properties of hardness and flexibility. The composition might well vary during growth. For example, tooth enamel in infants is mainly a protein matrix, whereas by adulthood enamel is approximately 90% hydroxyapatite. Since the biomineral structure is controlled by the cell promoting it, an unusual structure for the biomineral may result. For example, strontium sulfate normally crystallizes as flat rhombs, but in the plankton *Acantharia*, twenty (exactly!) $SrSO_4$ single crystals assume complex shapes. Calcium dominates the biomineral world, silicon and iron (see also Section 7.5) being the only other elements with even comparable roles. Silica, $SiO_2 \cdot nH_2O$, is present in the defensive spikes used against predators in certain plants and grasses (stinging nettles!). Various iron oxides occur in the teeth of chitons, limpets and beavers, and magnetite acts as a direction sensor in certain bacteria. Biomineralization is a topic of great interest to a variety of scientists including geologists, biologists, medical practioners and inorganic chemists. Attempts are being made to prepare synthetic materials which will simulate natural materials, e.g. bone substitutes for clinical use. The understanding of the underlying principles controlling biomineralization will be necessary to achieve this goal.

5.6 Health and the s-block

Table 5.5   Inorganic pharmaceuticals

Compound	One trade name	Value
Li carbonate or citrate	Camcolit	Used to treat both acute mania and prevent manic depressive bouts.
Mg salts	Magnesia	Antacids, laxatives.
Mg, Al hydroxides mixture	Maalox	Antacid. Long acting because relatively insoluble and thus slow base release.
$CaCO_3$ (mixed with Mg salts)	Nulacin	Antacid and antidiarroeal. Treatment of peptic ulcers.
$BaSO_4$	Baridol	X-Ray diagnosis of stomach and intestinal abnormalities.

Clinical trials have been promising for a material injected as a paste into a bone fracture site. The material hardens within ten minutes and conforms to the contours of the fracture. The material is a carbonated apatite called dahlite. It is gradually replaced by fresh living bone. It was developed by materials scientists and surgeons and, if sucessful, will have great advantages over current methods for fracture treatment. The plaster cast may become a thing of the past!

Although pharmaceuticals are usually associated with organic compounds, inorganic compounds, often simple, have been used in the treatment of many ailments, in some cases since ancient times (Table 5.5). An inorganic compound like sodium chloride is often used as an excipient. In some cases the metal ion is the important component (e.g. Li^+). In other instances, the alkali or alkaline earth cations simply deliver the inorganic anion (e.g. CO_3^{2-}, OH^-). Occasionally the compound, *in toto*, is vital (e.g. $BaSO_4$).

6 The d-block – redox chemistry

6.1 Introduction

The d-block of elements (transition elements) are a family of metals with, generally, similar characteristics. They represent the transition between the highly electropositive s-block metals and the electronegative p-block non-metals. The first row dominates the biologically important transition metals. The electronic configurations are:

vanadium	(Ar core) $4s^2\,3d^3$	cobalt	(Ar core) $4s^2\,3d^7$
chromium	(Ar core) $4s^1\,3d^5$	nickel	(Ar core) $4s^2\,3d^8$
manganese	(Ar core) $4s^2\,3d^5$	copper	(Ar core) $4s^1\,3d^{10}$
iron	(Ar core) $4s^2\,3d^6$	zinc	(Ar core) $4s^2\,3d^{10}$

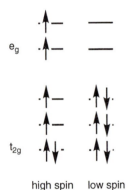

e_g

t_{2g}

high spin   low spin

Their behaviour as ions in solution is predominantly coordination and redox chemistry. The transition elements (particularly the middle members of the class) exhibit variable oxidation states. Iron, for example, can exist in oxidation states of II, III, IV and VI, although these are of variable stability. Iron (II) has the $3d^6$ electronic configuration. The octahedral arrangement of the six d electrons can be in two dispositions and this possible dichotomy plays an important role in iron–protein chemistry (Section 6.3). Antiferromagnetic coupling (Section 3.5) of two Fe(III) ions leads to diamagnetism. A pair of Fe(II) and Fe(III) ions coupled in this way behave like an isolated metal centre with a single unpaired electron and an EPR spectrum which exhibits three g values.

The transition elements are invariably present in biological material as coordination complexes, with ligands ranging from cellular components such as H_2O, small organic molecules and porphyrins, to the side chains of amino acids. All types of donor atoms can be coordinated by this versatile group of cations. The transition metal ions differ from the alkali and alkaline-earth metal ions in significant respects. They can usually be easily monitored. They tend to hydrolyse at physiological pHs, eqn 6.1.

$$Fe(H_2O)_6^{3+} \rightleftharpoons \text{polynuclear Fe(III) species (insoluble)} \qquad (6.1)$$

This hydrolysis must be prevented by metal ion chelation, a necessity in the storage and transport of transition metal ions in living systems (Section 7.5). Finally, unlike the s-block elements, they have a rich redox chemistry (except zinc). The important redox active metals in biochemistry are Fe (by far the leader), Cu and Mo. V, Co and Mn are also used, but much less often. Zinc, which is very important in other respects, exists only in the +2 oxidation state. The d-block metal ions can be associated both with small molecule binding and oxidoreductase enzyme activity (Section 6.3). Alternatively the metal centre may act *solely* as a conduit for electron transfer (Section 6.4). The latter apparently is the function of Cu in the blue (type-1) copper proteins, typified by plastocyanins (Section 4.2). In plants and certain algae, they transfer electrons from a cytochrome b_6f complex to P700* in photosynthesis (Section 6.4). In multisite metalloenzymes, one metal centre may function to shuttle electrons to another metal site where substrate binding and reaction occurs. Examples of this, which we will meet later, are cytochrome c oxidase, the FeMo protein in nitrogenase and probably the Mn_4 cluster in photosystem II. Redox and many other reactions take place in the cell in organelles (Fig. 6.1).

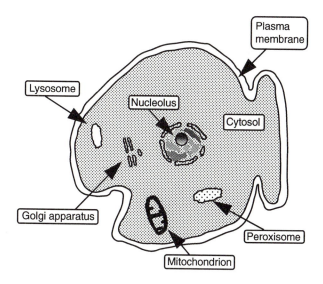

Fig. 6.1 A few organelles in animal cells. Included are lysosome (digestive enzymes location), peroxisome (important in liver cells for detoxification), mitochondrion (aerobic respiration, Fig. 6.5) and Golgi apparatus (manipulation of glyco- and lipoproteins). These organelles and the cytosol house important antioxidants (Section 8.6). Other organelles (not shown) aid in protein synthesis, as does the nucleus which fills a large part of the cell. Only a few of hundreds of different proteins in the mitochondrion are encoded by mitochondrial DNA. The rest are encoded by nuclear DNA, synthesized in the cytoplasm and then imported into the mitochondrion.

6.2 Electron donor and acceptor centres

There are a relatively limited number of types of moieties which act solely as electron donors and acceptors. These include cytochromes and other metal porphyrin-containing molecules and iron–sulfur centres. We will also encounter organic redox coenzymes in this section, because they often act in conjuction with metal centres.

Cytochromes

Cytochromes are heme proteins in which the electron transfer is typically between the Fe(II) and Fe(III) forms. Those termed a,b and c predominate and differ in the peripheral groups on the porphyrin or in the mode of attachment of the porphyrin to the protein (Table 6.1). We have already encountered porphyrin b (protoporphyrin IX) in Sections 2.1 and 6.2. The much less common cytochromes d have one or two dihydroreduced porphyrin rings and a

Table 6.1   Structures of cytochromes a–c

Type	Peripheral groups				
	2, 7,12	3	8	13,17	18
a	CH_3	$CH(OH)CH_2R^a$	$CH=CH_2$	$(CH_2)_2COOH$	CHO
b	CH_3	$CH=CH_2$	$CH=CH_2$	$(CH_2)_2COOH$	CH_3
c	CH_3	$CH(CH_3)S$ Cys - - Cys S$CH(CH_3)$		$(CH_2)_2COOH$	CH_3

$^{a)}$ R is $(CH_2CH=C(CH_3)CH_2)_3H$, a phytyl tail.

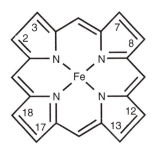

variety of peripheral groups. There has been a good deal of confusion in the nomenclature of cytochromes, arising from the haphazard history of their discoveries. Consider cytochrome f, a part of the cytochrome b_6f complex which functions in photosynthesis (Section 6.4), and got its name from *frons* (Lat. leaf). It is in fact a c-type cytochrome with the heme attached to two cysteines. The fifth ligand is his but the sixth is, surprisingly, a terminal peptide NH_2 group. In addition, the structure of cyt f is dominated by β-sheets and not α-helices, which are normally found in cytochromes. Numeric or maximum absorption wavelength (of the reduced cytochrome) subscripts are further subclassification ploys. Axial groups, typically met and his residues from the protein backbone, usually complete the octahedral coordination around the iron. This tight coordination prevents binding of small molecules and allows the entity to act solely as an electron carrier. Electron transfer involving the Fe(II) and Fe(III) states is facile since there is little structural change during the redox reaction (Section 6.4). Some enzymes are incorrectly referred to as cytochromes, e.g. cytochrome P-450. A cytochrome may also act in concert with one or more additional cofactor(s). Ubiquinol cyt c reductase (Section 6.4), for example, contains a diheme cyt b_1 and one cyt c, as well as an FeS cluster (see next section). The structures of several eukaryotic and prokaryotic cytochromes c (from horse, tuna and other species) show striking similarities in being relatively small and globular. There are few similarities among their amino acid sequences, but certain internal glycines and lysines near the heme and other functionally important residues are retained. A useful structure has obviously been conserved during evolutionary change.

Q. What advantages do porphyrin and corrin (see photosynthesis, Section 6.4) rings provide to biological systems?

Iron–sulfur centres

There are a large number of proteins containing a variety of FeS clusters, which have multifarious functions. The high-spin iron is coordinated to four sulfurs from inorganic sulfide and protein cysteines in an approximately

Table 6.2 Iron–sulfur compounds

Structure	Characteristics of Fe(s)	Occurrence
[Fe–0S] 	High-spin, slightly distorted tetrahedral Fe(II) and Fe(III). Oxidized form has $g = 4.3$, 9.0 values. $Fe(S\text{-}cys)_4^-$ Fe^{3+} $\uparrow\downarrow$ $Fe(S\text{-}cys)_4^{2-}$ Fe^{2+}	Rubredoxin EXAFS showed that all Fe–S bonds were similar in length, correcting an earlier X-ray structure.
[2Fe–2S] 	Antiferromagnetic coupling means (a) is diamagnetic and (b) is paramagnetic, and valence-localized (g=1.89, 1.95 and 2.05). $Fe_2S_2(S\text{-}cys)_4^{2-}$ $Fe^{3+}Fe^{3+}$ (a) $\uparrow\downarrow$ $Fe_2S_2(S\text{-}cys)_4^{3-}$ $Fe^{2+}Fe^{3+}$ (b)	Mainly in photosynthetic organisms.
[4Fe–4S] 	$Fe_4S_4(S\text{-}cys)_4^-$ $Fe^{2+}3Fe^{3+}$ (a) $\uparrow\downarrow$ $Fe_4S_4(S\text{-}cys)_4^{2-}$ $2Fe^{2+}2Fe^{3+}$ (b) $\uparrow\downarrow$ $Fe_4S_4(S\text{-}cys)_4^{3-}$ $3Fe^{2+}Fe^{3+}$ (c) Valence delocalized structures (a) and (c) are paramagnetic; (b) is diamagnetic.	HIPIP (*Chromatium vinosum*) shuttles between (a) and (b) and has E= +350mV, (c) can be obtained. Ferredoxin (*Clostridium aerogenes*) shuttles between (b) and (c) and has E~ −400mV. There are two [4Fe–4S] clusters.
[3Fe–4S] 	$Fe_3S_4(S\text{-}cys)_3^{2-}$ $[3Fe^{3+}]$ $\uparrow\downarrow$ $Fe_3S_4(S\text{-}cys)_3^{3-}$ $[Fe^{2+}2Fe^{3+}]$ In the oxidized form, there are three equivalent Fes and three g values at ~ 2.01. In the one electron reduced form, there is antiferromagnetic coupling between Fe centres.	Present in inactive form of aconitase. Activated by Fe^{2+} to complete the Fe_4S_4 cluster (Table 7.2).

tetrahedral arrangement around the metal ion. Several of these arrangements are known, as shown in Table 6.2, which includes the one iron–sulfur protein with only cysteine ligands. More than one cluster of the same or different type, as well as other redox active groups, may be present in one protein (e.g. hydrogenases, Section 8.3). They are particularly effective redox centres since they exhibit a range of potentials from approximately +350 mV to –600 mV. They are therefore found as components in many processes involving electron carrier activity, where they are termed *ferredoxins* (Fd), as well as in classical oxidoreductase enzymes. Until recently it was believed that FeS centres only had a redox role in biology, but they are now known to function in nonredox enzymes (Table 7.2) and to also have a structural role in the DNA repair enzyme, endonuclease III (Section 7.6). They therefore constitute one of the more important prosthetic groupings. MCD, EPR (*g* values) and Mössbauer (isomer shifts) spectroscopies have been very valuable in diagnosing the structures and assigning oxidation and spin states to Fe in the FeS proteins. Modelling has also been important, since the (almost too good to be true!) discovery that FeS clusters form spontaneously when certain iron and sulfur compounds are mixed.

Organic cofactors

Since several electrons are required in the reduction of certain substrates (e.g. four for O_2 and six for N_2) it is obviously necessary in some cases for there to be more than one redox centre in the protein to accomplish this. There are a limited number of oxidation states easily attainable with a particular transition metal, so that linking these with other (organic) redox centres aids in the stepwise transfer of electrons. Important organic cofactors are shown in Fig. 6.2. Ubiquinone (coenzyme Q) and plastoquinone function as electron acceptors in the electron-transport and photosynthetic chains respectively (Section 6.4). These quinones, as well as flavin mononucleotide (FMN) and flavin adenine dinucleotide (FAD), can form stable semiquinone radical species and $2e^-$ reduced hydroquinones (e.g. for ubiquinone, Fig. 6.2 and eqn 6.2). The $1e^-$ reduced semiquinones are important for the transfer of single electrons particularly when, as in complex III in the respiratory chain (Section 6.4), the primary source of electrons, namely nicotinamide adenine dinucleotide (NADH), is only capable of $2e^-$ transfer (eqn 6.3).

The protein itself may provide centres which can take part in one electron changes and act in concert with redox centres. Radicals derived from cysteine, tyrosine, tryptophan and glycine side chains in the protein have been implicated as intermediates in metalloenzyme mechanisms. Galactose oxidase catalyses the two electron oxidation of RCH_2OH to $RCHO$ by O_2 (Table 6.4). For some time it had been believed that the redox change during catalysis involved Cu(I) and Cu(III). The higher oxidation state of copper, although not common, has been well established in simple Cu(III) complexes. However, there is now substantial evidence that the oxidized active form of the enzyme contains Cu(II) and a tyrosyl radical (A in eqn 6.4). B is an intermediate in the proposed mechanism for the catalytic oxidation of RCH_2OH to $RCHO$ via the ketyl radical, $R^{\bullet}CHO^-$.

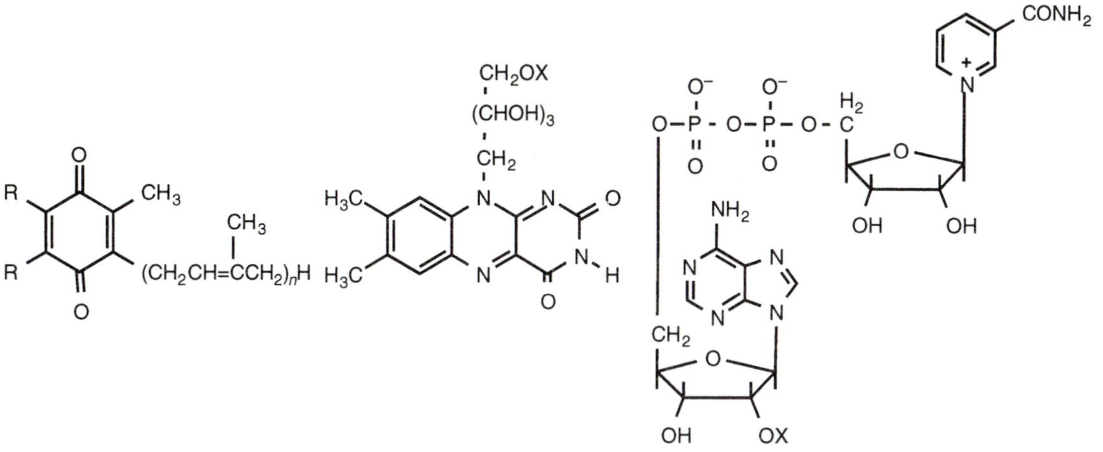

Fig. 6.2 Important organic cofactors (oxidized forms), often found tightly bound to proteins. Coenzyme Q (ubiquinone) **1**, R = OCH$_3$, n = 2–10, 10 in mammals, see Fig. 6.6. Plastoquinone, **1**,R = CH$_3$, n = 6–10, see Fig. 6.8. Vitamin K (phylloquinone), **1**, R = H, n = 4–7. The long tail allows for rapid diffusion in the lipid bilayer of the mitochondrial membrane. Flavin mononucleotide (FMN), **2**, X = PO$_3^{2-}$. Riboflavin, **2**, X = H. Flavin adenine dinucleotide, FAD, **2**, X = adenosine diphosphate. FAD and FMN are synthesized from FMN and riboflavin, respectively. Nicotinamide adenine dinucleotide, NAD$^+$, **3**, X = H, NADP$^+$, X = PO$_3^{2-}$.

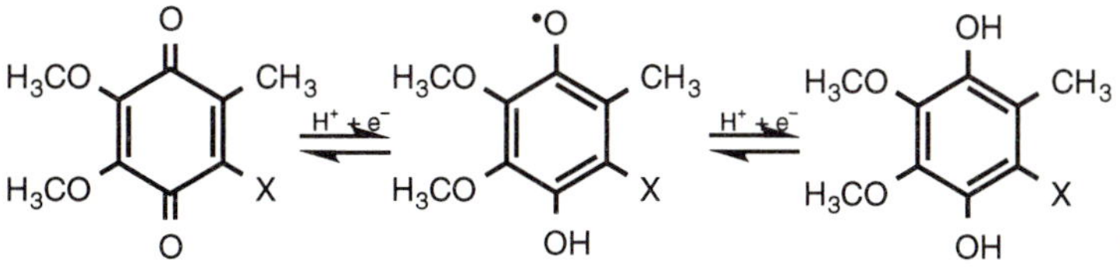

$$\text{NAD}^+ + 2e^- + \text{H}^+ \rightarrow \text{NADH} \tag{6.3}$$

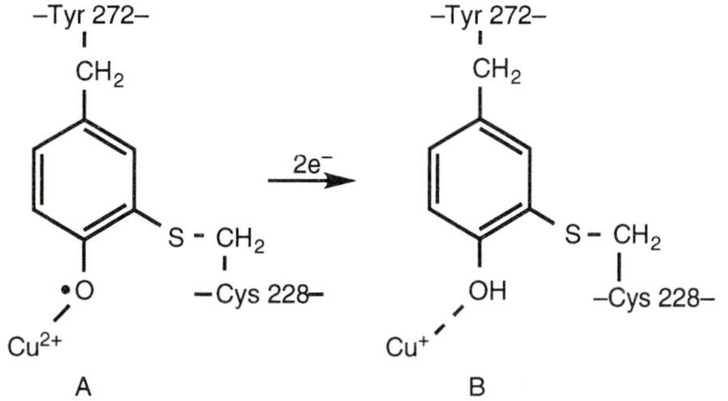

The appearance of the tyrosyl radical was first reported in the 1970s. It was observed in the ground state of the R2 subunit of *E. coli* ribonucleotide reductase. Its role in the action of the enzyme is still unclear but it is required for activity. The tyr radical cannot participate directly in H atom abstraction from the substrate as it is too far away. A cysteinyl radical may mediate electron transfer from the tyr radical.

Q. Why is the Cu(II) species A shown in Eqn. 6.4 likely to be EPR silent?

Pyruvate formate lyase catalyses a key step in anaerobic bacterial metabolism. The active form contains a stable glycyl radical which is required for activity. The unpaired electron is located on the α carbon of gly 734 giving rise to a simple EPR signal in the *E. coli* enzyme, in D_2O.

In general, in this book the electron associated with a free radical will not be shown. The tyrosyl radical also appears in photosystem II (Fig. 6.8). Evidence for the participation of tryptophan (cytochrome c peroxidase, Section 6.3), cysteine (mercuric reductase) and glycyl (pyruvate formate lyase) radicals in enzymic reactions has now been obtained. These radicals invariably operate in conjunction with other centres. The use of EPR spectroscopy has been vital in the detection of these radicals, which exhibit the characteristic simple $g = 2$ signal associated with a single unpaired electron.

6.3   Oxidoreductases

These comprise class 1 of the six enzyme classifications (Table 4.1). Many oxidoreductases contain heme iron, but non-heme iron, copper and molybdenum in several cluster arrangements are well represented. These metalloenzymes use O_2, O_2^- and O_2^{2-} as oxidants (oxidases) in aerobic organisms and SO_4^{2-} and other oxidants in anaerobic systems, which are much less well characterized. The hydrogenases use H_2 as the reductant (Section 8.3). Oxidoreductases act on a wide variety of substrates and a cofactor is often also employed. We have already encountered superoxide dismutase (Section 4.2) which catalyses the disproportionation of O_2^-, a reaction which can be considered as an intermolecular redox reaction.

Heme-containing enzymes

Table 6.3 shows a number of examples of oxidoreductases containing a heme centre, selected to illustrate the variety of sources, functions and structures exhibited by this versatile group. Most of the enzymes have had their structures solved by X-ray crystallography and in these cases their source is indicated.

Monooxygenases catalyse the insertion of a single oxygen atom from O_2 into a substrate and the reduction of the other O atom to water. Dioxygenases, in contrast, catalyse the insertion of both O_2 oxygen atoms into the substrate. Cytochrome P-450 (Table 6.3) is a very important example of a monooxygenase. It is responsible for catalysing the conversion of endogenous lipophilic bile acids and hormones, as well as exogenous drugs such as barbituates, codeine and others to water soluble products. These are more easily eliminated from the body because the conversion of C–H to C–OH has been effected. The crystal structure of bacterial P-450$_{cam}$ (the enzyme whose specific substrate is camphor) is known and the catalytic mechanism is well understood. We reproduce the catalytic cycle in Fig. 6.3 to show the importance of the oxidation and spin state of the iron and the general machinations of the metal centre and substrate which must occur in order to effect the transformations carried out by the active site. Although there are similar considerations for many other metal-containing oxidoreductases, the details will, of course, differ.

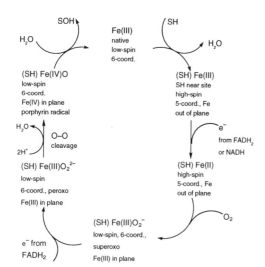

Fig. 6.3 Cycle for cytochrome P-450 catalysed oxidation of SH to SOH using O_2. The active oxygenation species is an oxyferryl complex, probably Fe(IV)O and a porphyrin cation radical. Fe represents protoporphyrin IX Fe(Scys)(H_2O), see Fig. 2.3.

Table 6.3   Selected metallooxidoreductases (heme containing)

Cytochrome P-450 :(*Pseudomonas putida*) P-450$_{cam}$ is the bacterial form which catalyses the hydroxylation of D–camphor to 5-*exo*-hydroxycamphor and whose structure is known. Low-spin Fe(III) heme b with axial cys and H_2O (Fig. 2.3). For mechanism see Fig. 6.3. One of a large family of enzymes catalysing hydroxylations at C, N and S centres, e.g.:

$$\overset{\displaystyle |}{\underset{\displaystyle |}{-C}}- H \longrightarrow \overset{\displaystyle |}{\underset{\displaystyle |}{-C}}- OH$$

Ubiquitous from bacteria to man. Cytochromes P-450 are terminal components of the nonphosphorylating electron transport chain in adrenal mitochondria and liver microsomes.

Peroxidase: (horseradish) high-spin Fe(III) heme b with axial his and H_2O. Two Ca^{2+} ions confer stability.For mechanism see eqns 6.6 and 6.7. Mn- and V- containing peroxidases are also known.

$$H_2O_2 \ + \ SH_2 \ \rightarrow \ 2H_2O \ + \ S$$

Peroxidases are present in plants and animals (spleen and lung in mammals) acting on peroxides in the presence of SH_2 (e.g. toxins).

Catalase: (bovine liver) high-spin Fe(III) heme b with axial tyr and H_2O. For mechanism see eqns 6.6 and 6.7. A binuclear Mn form is also known.

$$2H_2O_2 \ \rightarrow \ 2H_2O \ + \ O_2$$

All aerobic tissues contain catalases which catalyse the breakdown of H_2O_2.

Cytochrome c oxidase: (bovine) isolated six-coordinate heme a and Cu_A^{2+} sites and a coupled five-coordinate heme a$_3$, Cu_B^{2+} pair. The Cu_A site comprises two Cu ions with bridging cysteines (see Sections 3.1, 6.4 and 8.2). Mg^{2+} and Zn^{2+} have unknown roles.

$$4cyt \ c \ (Fe^{2+}) \ + \ 4H^+ \ + \ O_2 \ \rightarrow \ 4cyt \ c \ (Fe^{3+}) \ + \ 2H_2O$$

Complicated enzyme which catalyses the terminal transfer of electrons to O_2.

The peroxidases and catalases are extremely important for the removal of toxic H_2O_2, and in one case for the generation of a weapon (Fig. 6.4). Since the heme b cofactor (Table 6.1) predominates in many of these enzymes, it is encumbant upon us to try to explain the different behaviours exhibited by the numerous iron proteins which contain it.

Q. What is the likely result of the reaction of the resting form of P-450 with a peroxide (this is called the peroxide shunt, see Fig. 6.3)?

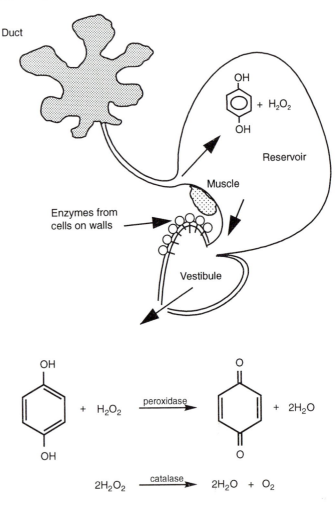

Fig. 6.4 Gland of the bombardier beetle. When desired the reservoir fluid is squeezed into the vestibule *via* a muscle controlled valve. In the vestibule, enzymes from cells on the wall catalyse the rapid reactions shown, to produce quinones (toxic to foes) and O_2 and H_2O as steam (exothermic reactions), a mixture which is squirted through the abdominal tip in any direction rapidly and repeatedly. This constitutes a very effective defensive chemical spray.

The fate of oxygen in the presence of an Fe heme centre

If the iron heme centre (PFe) has two strong axial ligands completing the octahedron around the Fe, it is likely that this entity will be insensitive to oxidation by O_2 and it may serve solely as an electron carrier (Section 6.4). If however, only one protein amino acid (e.g. his) is attached to the iron, and the sixth position is vacant, or perhaps occupied by labile (Section 3.3) H_2O, then O_2 can rapidly bind to the PFe(II) entity at that site. Subsequent events will be strongly influenced by the environment of the iron, provided by the protein, in ways which are still incompletely understood. If, for example, the pocket in which the heme sits is strongly hydrophobic, it will then be protected from access by hydrophilic substances (e.g. anions) and the PFe(II)–O_2 entity is likely to be relatively stable. Such a protein can function as a storer or transporter of O_2, e.g. myoglobin and hemoglobin (Section 8.7). If on the other hand reduction (and protonation) of the O_2 in the PFe–O_2 complex cannot be prevented, then a ferric hydroperoxy compound Fe(III)–O_2H^- is formed. This undergoes heterolytic O–O bond cleavage to produce a very reactive species, probably $^\bullet$PFe(IV)O (eqn 6.5), in which iron is in the +4 oxidation state and the porphyrin has been one-electron oxidized, as in for example P-450, Fig. 6.3. A similar species might be expected, and does indeed arise, in the reactions of the PFe(III) moiety in peroxidase and catalase, with H_2O_2 (eqn 6.6). Heterolytic cleavage of the coordinated O–O moiety is aided by electron donation from the electron rich cys axial ligand in P-450 and by judicious placement of polar residues near the Fe–O sites in peroxidase and catalase (Fig. 4.3). What now happens to the $^\bullet$PFe(IV)O species? If the substrate binds close to the metal site, as has been shown to be the case with P-450$_\text{cam}$, *oxygen atom transfer* from the $^\bullet$PFe(IV)O to the bound substrate can occur and the resting form of the enzyme, PFe(III) is reformed (eqn 6.5). With catalase and peroxidase however, the structural details around the heme site, particularly its being highly polar, prevent substrate, usually lipophilic, from binding close to the Fe. Therefore, only electron transfer (two sequential steps) occurs in forming the final product (eqns 6.6 and 6.7).

$$(SH)PFe^{III} + O_2 + 2e^- + 2H^+ \rightarrow\rightarrow\rightarrow (SH)^\bullet PFe^{IV}O + H_2O$$

$$(SH)^\bullet PFe^{IV}O \rightarrow PFe^{III} + SOH \qquad (6.5)$$

Production and fate of $^\bullet$PFe^{IV}O species in P-450

$$PFe^{III} + H_2O_2 \rightarrow PFe^{III}(H_2O_2) \rightarrow {}^\bullet PFe^{IV}O + H_2O$$

$$\underset{\text{compound I}}{{}^\bullet PFe^{IV}O} + SH_2 \rightarrow \underset{\text{compound II}}{PFe^{IV}O} + {}^\bullet SH + H^+ \qquad (6.6)$$

$$PFe^{IV}O + {}^\bullet SH + H^+ \rightarrow PFe^{III} + S + H_2O \qquad (6.7)$$

Production and fate of $^\bullet$PFe^{IV}O species in peroxidase and catalase ($SH_2 = H_2O_2$)

The Fe coordination in myoglobin and peroxidase is quite similar and both have an axial imidazole group. Thus other factors must determine their different reactivities. One of these may be the lack of appropriate amino acids in myoglobin to aid heterolytic O–O bond cleavage as occurs with peroxidase.

The initial product of the reaction of yeast cytochrome c peroxidase with H_2O_2 is PFe(IV)O and trp 191 (not porphyrin) radical. The amino acid radical was identified by site directed mutagenesis and electron nuclear double resonance (ENDOR) spectroscopy. The latter is an extension of the EPR method which allows determination of the hyperfine coupling constant, often unresolvable by EPR alone.

Q. What properties (compared with other heme proteins) is the mutant ala80met cyt c likely to have?

Rapid scanning after rapidly mixing PFe(III) and H_2O_2 enables the spectrum of compound I to be measured and its structure assessed. Compound I is only stable for a few hundred milliseconds.

Non-heme containing enzymes

Iron is well represented in non-heme enzymes, but cobalt, copper, manganese and molybdenum also feature in these. Table 6.4 illustrates several important oxidoreductases.

There are three types of copper centres in proteins. Type 1 is present in plastocyanin (Section 4.2). Type 2 Cu sites have normal εs and EPR signals. Type 3 Cu sites, found in oxyhemocyanin (Section 3.5), have high εs at ~ 330 nm and antiferromagnetic coupling of a pair of Cu ions (Section 3.2). All three types of Cu site occur in ascorbate oxidase and laccase, in which the trinuclear Cu cluster was first defined by MCD (Table 3.1).

All Mo-containing enzymes, except nitrogenase, are oxotransferases which catalyse oxo group transfer to or from substrates. The Mo centre contains a pterin cofactor:

Oxidation states of VI, V and IV are used by Mo.

Q. What do you predict would be the properties of a SOD from which only the Zn had been removed? How would you prepare such a species?

Table 6.4   Selected metallooxidoreductases (non-heme)

Protocatechuate-3,4-dioxygenase: from *Pseudomonas aeroginosa* is a large oligomer which contains high-spin non-heme trigonal bipyramidal Fe(III) (tyr, his, H_2O; his, tyr). Red Fe(III)–tyr LMCT band. It is an intradiol dioxygenase. Coordination of the catechol by Fe activates it for attack by O_2 and subsequent reaction.

Superoxide dismutase: (bovine erythrocytes) CuZn form in which Cu^{2+} (4 his) and Zn^{2+} (3 his, asp) share a deprotonated imidazolate from his. Mechanism involves Cu(II) and Cu(I) reactions with O_2^-. Almost all eukaryotes contain a CuZn form. Fe- and Mn-containing forms also known, but these are structurally unrelated to CuZnSOD.

$$2O_2^- + 2H^+ \rightarrow O_2 + H_2O_2$$

Ubiquitous among aerobic organisms. Protects them from the mutagenic superoxide radical.

Galactose oxidase: (*Dactylium dendroides*) Type 2 Cu(II) square pyramidal centre (2 his, tyr, $CH_3CO_2^-$; tyr). For mechanism see eqn 6.4.

$$RR'CHOH + O_2 \rightarrow RR'CO + H_2O_2$$

Responsible for the oxidation of alcohols in fungi.

Xanthine oxidase: no crystal structure. EXAFS and other techniques indicate Mo pterin, FAD and 2[2Fe–2S] in each subunit of the dimer. Mechanism involves cycling of Mo between +4 and +6 oxidation states.

hypoxanthine   xanthine   uric acid

In mammals in liver and intestines. The enzyme is a target of drugs for gout sufferers (caused by precipitation of sodium urate).

6.4 Electron movement in metalloproteins

We have already examined the extremely important group of oxidoreductases which catalyse electron transfer reactions. In addition, electron movement is required in such different processes as respiration, photosynthesis and nitrogen fixation, and in these, redox centres pass on electrons to neighbours but are not themselves involved with substrates. The transfer of electrons is sometimes coupled with proton transfer which may then be used in the synthesis of ATP. The movement of electrons in biological systems can occur over large distances, as much as 15 Å or more.

There is extensive use of a variety of electron transfer proteins in (a) the last stage of the catabolism of carbohydrates (the electron transport chain) and in (b) the initial stage of the photosynthetic production of carbohydrates (the light dependent stage, eqn 6.8). In (a) CO_2, as well as reducing equivalents (e^-) in the form of NADH and $FADH_2$, arise from the enzymatically catalysed breakdown of carbohydrates by H_2O. The electrons are then transferred through a chain to finally convert O_2 and H^+ into H_2O. In (b), by contrast, H_2O is converted into O_2, H^+ and electrons via a photoexcited centre. The electrons are then transferred stepwise to produce NADPH, the reducing power of which is used to convert CO_2 into carbohydrates in the later (light independent) steps. The electron transfer chains in (a) and (b) also fulfil an important role in aiding in the enzymatically catalysed synthesis of ATP from ADP. The energy released (exergonic) in the hydrolysis of ATP can be used to drive reactions requiring the input of energy (endergonic).

> NADH is used primarily in the electron transport chain. NADPH is used generally for reductive biosynthesis in photosynthesis.

$$O_2 + 4H^+ + 4e^- \xrightleftharpoons[\text{light dependent photosynthesis}]{\text{electron transport chain}} 2H_2O \qquad (6.8)$$

Electron transport chain

The large amount of energy (2.9×10^3 kJmol^{-1}) which is released in the catabolism of glucose:

$$C_6H_{12}O_6 + 6O_2 \rightarrow 6CO_2 + 6H_2O \qquad (6.9)$$

must be controlled to be of value. This is achieved by releasing the energy in a large number of steps. In a series of enzymatically catalysed reactions which take place in the cytosol (Fig. 6.1) and mitochondrial matrix (Fig. 6.5) in three stages (glycolysis, link reaction and Krebs cycle), the carbohydrate is broken down into carbon dioxide and reduced coenzymes (mainly NADH, but some $FADH_2$). What concerns us is the fourth and final stage which takes place on the inner surface of the inner membrane of the mitochondrion, and in which oxygen is used to oxidize the reduced coenzymes (eqns 6.10 and 6.11).

> Q. Why should the release of energy be controlled?

$$2NADH + O_2 + 2H^+ \rightarrow 2NAD^+ + 2H_2O \text{ (6ATP produced)} \quad (6.10)$$
$$2FADH_2 + O_2 \rightarrow 2FAD + 2H_2O \text{ (4ATP produced)} \qquad (6.11)$$

Electron transfer chain segment:

cyt c_1Fe(III) + e^- → cyt c_1Fe(II)

cyt c_1Fe(II) + cyt cFe(III) →

cyt c_1Fe(III) + cyt cFe(II)

cyt c_1Fe(III) + e^- → etc.

The simplicity of these equations disguises the complexity of the system (Fig. 6.6). The numerous steps, many involving metalloproteins and the cofactors shown in Fig. 6.2, serve to keep the reactants well separated, as well as dividing the energy associated with the reactions into small portions. Reducing equivalents are passed down the chain (indicated by a bold arrow) from the reduced coenzymes to the terminal O_2. At various points along the chain (indicated by asterisks in Fig. 6.6) there is a transfer of protons from the matrix to the intermembrane space (e.g. by the process shown in eqn 6.2). Protons then flood back through the membrane using an embedded ATPsynthase to initiate ATP synthesis (eqn 6.12 and Section 8.10), the large majority of which is produced in this chain.

$$ADP + P_i \rightarrow ATP + H_2O \qquad (6.12)$$

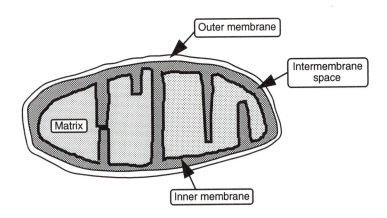

Fig. 6.5 Rough sketch of mitochiondrion. This is an organelle found in the cytoplasm of all eukaryotic cells. The outer, lipid bilayer membrane is easily permeable. The inner membrane is rich in proteins and not easily permeable. It contains the electron transport proteins and ATP synthase. The matrix contains the genetic machinery.

$E^{o\prime}$ is the standard reduction potential at pH = 7.0, which is the biological standard state. The value will differ from the standard reduction potential E^o (for which $2H^+ + 2e^- \leftrightarrow H_2 = 0$ at pH = 0.0), if H^+ is involved in the half reaction. At pH = 7.0 therefore, NADH will reduce Q.

Complexes I–IV are mobile within the membrane and, as would be anticipated, there are sucessively increasing reduction potentials for the participants from left to right in Fig. 6.6. For example, for the couples NAD^+ / NADH, Q / QH_2 and O_2 / H_2O, the values of $E^{o\prime}$ are, respectively, -0.32, $+0.10$ and $+0.82$ volts. Crucial electron carriers in the chain are cytochrome c and cytochrome c oxidase (complex IV, Table 6.3). Cytochrome c has two axial ligands which emanate from the protein backbone (rather than a vacant coordination site or a labile H_2O, Section 6.2) and a porphyrin edge which is exposed to solvent. Its primary role is to shuttle electrons by alternately binding between the complexes III via cyt c_1, and IV via $(Cu_A)_2$ and cyt a. In IV, stepwise transfer of $4e^-$ to the cyt a_3–Cu_B binuclear complex where O_2 is bound, leads to its reduction to H_2O (eqn 6.8). Cytochrome c oxidase is the terminal component of the respiratory chain in all animals, plants, yeasts and some bacteria. At low O_2, *E. coli* use fumarate or nitrate as a terminal oxidant and a transcription factor (FNR) operates.

FNR contains an Fe_4S_4 cluster, which is unstable in O_2, thus rendering FNR ineffective.

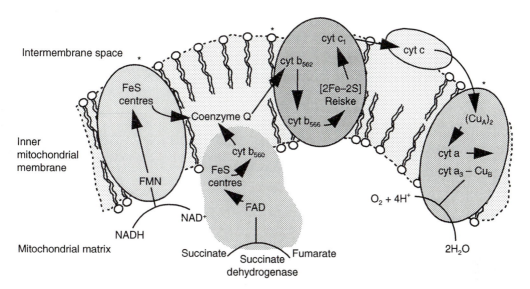

Fig. 6.6 Simplified diagram of the respiratory chain in a mitochondrial membrane showing the sequence of complexes which carry the electrons (direction shown by bold arrows) from NADH to O_2. At the points indicated by an asterisk protons are pumped. Other proteins are also present, e.g. 4–6 small subunits in complex III but their functions are unknown. Complexes I–IV from left to right.

Photosynthesis

Photosynthesis operates in the membrane systems of green plants and photosynthetic bacteria. The overall reaction, eqn 6.13 can be seen to be the reverse of oxidative catabolism (eqn 6.9), although the mechanistic details differ. Photosynthesis takes place in the chloroplasts (Fig. 6.7) found in the

Q. Why are the chains in Figs 6.6 and 6.8 also referred to as oxidative phosphorylation and photophosphorylation?

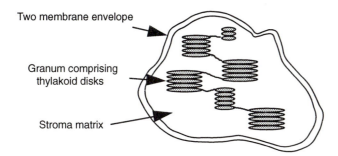

Fig. 6.7 Rough sketch of a chloroplast in green plant cells. This organelle is found in the cytoplasm of cells of all green plants, which also contains organelles found in animal cells (Fig. 6.1). The thylakoid disks or membrane enclose an internal space (lumen). They are continuous and highly folded. Both PSI and PSII are embedded in the membrane. The stroma matrix contains concentrated solutions of DNA, RNA and enzymes for carbohydrate synthesis in the dark reaction.

cytoplasm of cells in all green plants. There are two components to photosynthesis: a light-dependent stage (light reaction) occuring in the thylakoid discs which make up each granum, and a light-independent stage (dark reaction) which takes place in the stroma matrix, and makes use of the products of the light reaction.

$$6CO_2 + 6H_2O \xrightarrow{h\nu} C_6H_{12}O_6 + 6O_2 \qquad (6.13)$$

Light reaction

Our present understanding of the complicated process is summarized in Fig. 6.8. There are two photosystems, termed photosystem I (PSI) and II (PSII).

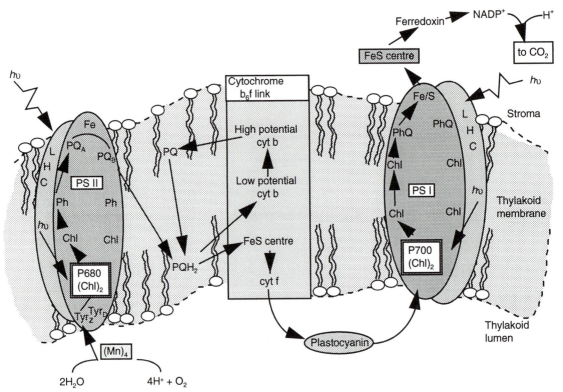

Fig. 6.8 Simplified representation of the electron transport chain in thylakoid membranes (photosynthesis). The two photosystems I and II operate in tandem to extract electrons and protons from H_2O and direct them (flow indicated by arrows) to carbon dioxide.

Certain anaerobic green and purple photosynthetic bacteria use a variant of either PSI or PSII. Aerobic cyanobacteria and plants use PSI linked to PSII, although there is good evidence for significant by-passing of PSI (Fig. 6.8). Let us consider plants. Both PSII and PSI use a 'special pair' of chlorophyll, $(Chl)_2$, molecules as the primary electron donor (P_D) and a chlorophyll or pheophytin (Ph), Fig. 6.9, as the primary electron acceptor (P_A). These are the targets of sunlight which has been transmitted through light harvesting

centres (LHC). On excitation by light a radical pair is formed in the reaction
centre within picoseconds (eqn 6.14):

$$P_DP_A \xrightarrow{h\nu} P_D^*P_A \xrightarrow{\text{internal e}^- \text{ transfer}} P_D^+P_A^- \qquad (6.14)$$

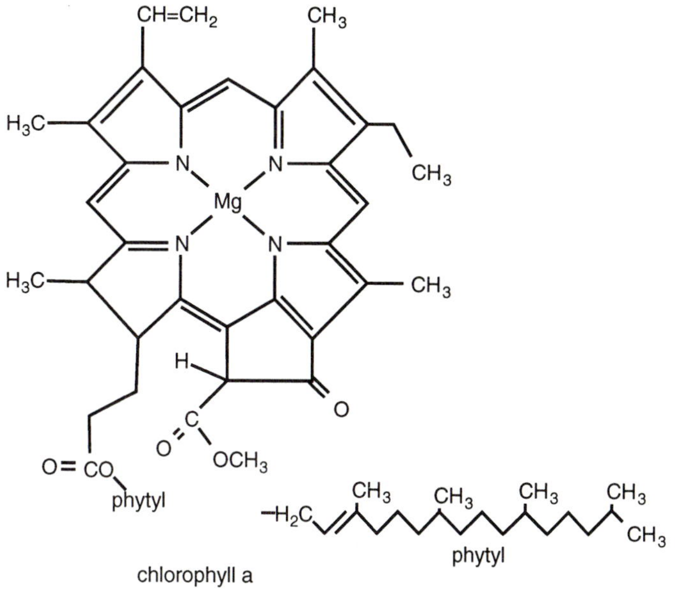

chlorophyll a

LHCs consist of many antenna chlorophylls which pass the energy of absorbed photons by exciton transfer (in less than picoseconds) from molecule to molecule until the reaction centre is reached. They also contain (fewer) accessory pigments, such as carotenoids, which absorb at wavelengths not covered by chlorophylls. LHCs also regulate the fluctuating intensities of sunlight. About half of the chlorophyll on earth is bound in LHCs. Bacterial and plant LHC II differ structurally.

Fig. 6.9 Chlorophyll a. Chlorophyll b and bacteriochlorophylls a and b (used in bacteria) are variations of chlorophyll a. Pheophytins (Ph in PSII) are Mg free chlorophylls. These and carotenoids allow a broad range of light absorption.

The remainder of the scheme is concerned with (a) the *rapid transfer* of an electron from P_A^- to a chain of electron acceptors (just as in the respiratory chain) and (b) the *rapid uptake* of an electron by P_D^+ from a chain of electron donors. In PSII, P_D^+ is a powerful oxidant in the form P680+ ($E^\circ \sim 1.1$ volts). It absorbs electrons released stepwise from the oxidation of H_2O by a Mn cluster (in mixed oxidation states involving Mn (II) through Mn(IV)) with the evolution of O_2 gas *via* tyr$_z$ radical (eqn 6.15). The relatively weak oxidant P700+ ($E^\circ \sim 0.4$ volts) produced by light excitation in PSI takes up electrons originating in P680− and terminating in reduced plastocyanin, the

Q. Why might there be branching at the free plastoquinol point?

Q. What are the advantages of using Mg^{2+} as the metal ion in chlorophyll?

$$2H_2O \rightarrow 4H^+ + O_2 + 4e^- \qquad (6.15)$$

last member of the chain emanating from PSII via the cytochrome b$_6$f link. The strong reductant concommitantly produced, P700−, transfers electrons in PSI to produce finally NADPH, which goes on to reduce CO_2 in the dark reaction. The passage of electrons is indicated by bold arrows in Fig. 6.8. There are duplicated tyr, Chl and Ph in PSII and Chl and PhQ (phylloquinones) in PSI. These are, quite surprisingly, not used in electron transfer. Both bound plastoquinones (PQ$_A$ and PQ$_B$) are used in PSII and pass electrons to mobile PQ($\rightarrow$ PQH$_2$). As in the mitochondrial chain, organic cofactors are used (Fig. 6.2) and there is a translocation of protons, now from thylakoid to stroma, which through the agency of ATP synthase leads to the production of ATP.

Our understanding of photosynthesis has been aided considerably by the recent acquisition of pure material for spectroscopic studies and by the X-ray investigations of the purple bacterium *Rhodopseudomonas viridis* reaction centre as a model for PSII. It is clear that bioinorganic species containing Mg, Mn, Fe or Cu are well represented in the electron transfer centres of PSI and PSII.

Dark reaction

Carbon dioxide and water are converted into glucose in the dark reaction, employing NADPH and ATP, in a series of steps termed the Calvin cycle. The dark reaction in eqn 6.16,

$$12NADPH + 12 H^+ + 6CO_2 \rightarrow C_6H_{12}O_6 + 6H_2O + 12NADP^+ \tag{6.16}$$

combined with the overall result of the light reaction eqn 6.17 (which is a combination of eqns 6.3 and 6.15), is easily seen to lead to eqn 6.13. ATP

$$2NADP^+ + 2H_2O \xrightarrow{\ h\nu\ } 2NADPH + O_2 + 2H^+ \tag{6.17}$$

generated in eqn 6.17 is used in eqn 6.16.

Rates of electron transfer

In a one electron change, $-\Delta G^\circ = FE^\circ$, where F is the Faraday constant and E° is the standard reduction potential. A reaction is favourable if $-\Delta G^\circ$ or E° is positive.

We have so far not considered in detail the rates of electron transfer reactions although they are obviously very important. The rate at which an electron is transferred from one redox centre to another has been determined experimentally and assessed theoretically in a large number of chemical systems. An understanding of the phenomenon in simple metal complexes, and later in specially designed organic molecules, laid the foundation for the application of the theory to biological systems.

Electron transfer may be *between* two molecules (intermolecular) or *within* a single molecule (intramolecular). Because in an intermolecular reaction there is necessarily some interaction between two molecules prior to electron transfer (the quantitative aspects of which are not always easy to assess) it is somewhat easier theoretically to examine intramolecular electron transfer. The important parameters which affect rates of electron transfer are:

(a) the potential drive underlying the electron transfer (free energy of reaction, $-\Delta G^\circ$);

(b) the reorganization energies (λ) required to have bond lengths and extents of solvation in the oxidised and reduced centres equalized in the transition state so that electron transfer can occur;

(c) the distance (D) between the redox centres.

In practice, this is the distance between the centres of the atoms at the edges of the donor and acceptor moieties. In addition, the parameter D will be modified by a term β, which is a measure of the effectiveness of the medium in the electronic coupling of donor and acceptor. The shortest possible distance will be at van der Waals contact ($D_o = 3.6$ Å).

An expression relating these parameters has been derived in which the rate constant (k, s^{-1}) for intramolecular electron transfer is given by:

$$k = A \exp\left[-\beta(D - D_o)\right]\left[-(\Delta G^\circ + \lambda)^2 \,/\, 4\lambda RT\right] \qquad (6.18)$$

where A is a constant, $\sim 10^{13}$. Examination of eqn 6.18 shows that for a particular reaction pair (constant D) the value of k will increase as $-\Delta G^\circ$ increases, reaching a maximum (k_{max}) when $-\Delta G^\circ = \lambda$. Thereafter, as $-\Delta G^\circ$ increases ($>\lambda$), the value of k will decrease. The prediction of this so called 'inverted region' which intuitively seems so unlikely, has been confirmed in several systems and has led to universal acceptance of the fundamental concept embodied in eqn 6.18.

Let us apply this concept to a biological system in which there are a number of different pairs of donor and acceptor sites and each pair is separated by a different distance. We can use the system to experimentally test the linear relationship which exists (eqn 6.18) between log k_{max} and D, k_{max} being used so as to correct for the ΔG° and λ terms, i.e. eliminate the term containing them. The photosynthetic reaction centre from the bacterium *Rhodopseudomonas viridis* (see above) is one such system. Distances between several of the redox sites in the system and the attendant k_{max} values for electron transfer are known. For example, there is a physiologically important electron transfer from photoreduced bacteropheophytin (BPh$^-$) to a quinone Q_A site over a distance of 9.6 Å (compare PSII in Fig. 6.8). If the native quinone is replaced by a quinone with a different E° value, this will impose a different ΔG° value. If the rate constants are measured for electron transfer from BPh$^-$ to the various quinone derivatives, a value of k_{max} for that particular transfer and distance D can be calculated using eqn 6.18. Figure 6.10 shows the plot of log k_{max} vs D for electron transfer between a number of redox pairs within the photosynthetic centre of *Rp. viridis*. An excellent linear plot is obtained extending over a 20 Å difference in the electron path distance. The value of A is 10^{13} s^{-1}, the rate constant at van der Waals contact and the value of β is 1.4 Å^{-1}. A number of other systems have also been examined. These include cytochrome c and azurin in which modifications have been made to enable electron transfer to be measured between the metal site and specific amino acid residues within the molecule. Although some individual values of k_{max} vs D lie reasonably well on the line of Fig. 6.10, a much lower value (1.1 Å^{-1}) for the slope ($-\beta$) is observed for the modified cytochrome c and azurin. This lower value for the β factor means that donor-acceptor electronic coupling decays much more slowly when $\beta-$strands separate the reaction centres (as is largely the case in azurin) than when α-helices predominate as in the photosynthetic reaction centre. Therefore, although an approximately exponential relationship holds between rate constant and distance for electron transfer within many biological materials, the nature of the medium plays an important role, and this is being actively investigated.

$$\log k_{max} = \log A - \frac{\beta(D - 3.6)}{2.303}$$

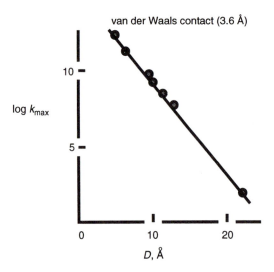

Fig. 6.10 Log of the optimized rate constant (k_{max}) vs edge-to-edge distance, D, for electron transfer between a variety of sites in the photosynthetic reaction centre from the bacterium *Rhodopseudomonas viridis*.

We now return to the chain reactions of the previous sections. From Fig. 6.10 there is roughly a tenfold change in rate for a 1.7 Å change in distance. This means that the rates of electron transfer across the 25–30 Å distances which separate some of the respiratory chain complexes would be much too slow (half-lives of seconds to days) to be of practical value. The difficulty is obviated by using a series of intervening redox centres. The 'inverted region' predicted by eqn 6.18 above may be important in photosynthesis. The initial charge separation (see eqn 6.14) $P_D^+ P_A^-$, which is the precursor to the productive electron transfer chain, is maintained in spite of a competing and non-productive charge recombination back to the ground state. This is surprising in view of the similar distances involved and a large $-\Delta G^\circ$ value associated with the latter process. It may well be that the large $-\Delta G^\circ$ value forces the charge recombination reaction into the 'inverted region' with an attendant slower rate. In this way the rate of charge recombination is less competitive with the rate of charge separation.

7 The d-block — nonredox chemistry

7.1 Introduction

The transition elements with a propensity for facile one or two electron changes, namely Fe, Cu and Mo, feature heavily in the oxidoreductases. The other transition elements, particularly Mn^{2+} and Zn^{2+} with stable d^5 and d^{10} configurations, respectively, but also Ni^{2+}, are important as metal centres in enzymes which catalyse nonredox reactions. The coordinated ligands emanate mainly from the protein (Section 2.1) but a coordinated water molecule is also usually present. The H_2O is labile when attached to a first row transition metal bivalent ion, half-lives of exchange being in the milli- to the microsecond range. This lability is essential for the metal ions to operate in metalloenzymes, since substitution of H_2O by substrate is often a necessary step and is rarely rate limiting in the overall mechanism. In some cases, the substrate will add to the existing coordinated ligands. This is possible because these metal ions have variable coordination numbers, usually from 4 to 6. Lability is lost when the metal ion is strongly chelated, as is required when the purpose of the ligand is to help transport the metal ion (Section 7.5).

7.2 Metalloenzymes

The first row transition elements (especially Zn^{2+}) are well represented in the hydrolase, lyase and isomerase classes (Table 4.1) of enzymes. Selected examples are shown in Tables 7.1 and 7.2. Zinc is essential (with Mg^{2+}) in the nucleotidyl transferases, including the DNA and RNA polymerases. In these enzymes the M^{2+} ion (often part of a binuclear metal cluster) provides a nucleophilic OH^- ion in the form of M^{2+}–OH^-. This attacks a phosphoryl or carbonyl centre, which may be coordinated to the metal ion at some time during the process. Hydrolysis or hydration is assisted by amino acids near the metal site. The M^{2+} also stabilizes intrinsic (reactants) or developing (five-coordinate intermediates) negative charge and may organize the reactants spacially for easier interaction. *In toto*, a not insignificant contribution!

   As with the entries in Tables 6.3 and 6.4, most of the metalloenzymes in Tables 7.1 and 7.2 have known crystal structures and their sources are indicated. We shall discuss, in some detail, two systems as typifying the sorts of reactions and behaviours that are likely to be encountered with these metalloenzymes or metallocoenzymes.

Table 7.1   Selected hydrolases

Q. How would you rationalize the requirement of CO_2 for *in vitro* assembly of the urease Ni centre?

- *Alkaline phosphatase (E. coli)*: two Zn^{2+} ($1Mg^{2+}$ also), operating at an optimum pH of ~ 8.
- *Purple acid phosphatase*: non-heme binuclear $Fe^{2+}Fe^{3+}$ (mammalian) or $Zn^{2+}Fe^{3+}$ (plant), operating at an optimum pH of 4–6.

$$RO-\overset{\overset{\displaystyle O}{\|}}{\underset{\underset{\displaystyle O^-}{|}}{P}}-O^- \xrightarrow{H_2O} ROH \; + \; HO-\overset{\overset{\displaystyle O}{\|}}{\underset{\underset{\displaystyle O^-}{|}}{P}}-O^-$$

Both phosphatases operate *via* a Zn–OH⁻ or Fe–OH⁻ species in a binuclear site which binds substrate and covalent phosphorylated enzyme intermediates ($E–PO_3^{2-}$). Hydrolysis of ($E–PO_3^{2-}$) yields a bridged phosphate which then dissociates.

- *Urease (Klebsiella aerogenes)*: binuclear Ni^{2+} (three- and five-coordinated) with an unusual bridge (carbamate formed from a lys ε-amino residue). First enzyme shown to contain nickel. Catalyses the degradation of urea at about 10^{14} times the rate of the uncatalysed reaction (compare Fig. 4.7).

$$(NH_2)_2C = O \xrightarrow{H_2O} 2NH_3 + CO_2$$

Urea is the end product of metabolism. Bacterial ureases are implicated in infection-induced urinary stones.

- *β-Lactamase II* (penicillinase) (*Bacillus cereus*): one class (B) contains tetrahedral Zn^{2+} (3 his, cys) at the catalytic centre. Three other classes are non-metallo serine hydrolases.

penicillin   penicillinoic acid

β-Lactamase mediated resistance is a clinical problem encountered in using most β-lactams, including penicillin and cephalosporins.

- *Carboxypeptidase A* (bovine): a thoroughly studied Zn^{2+} (2 his, glu and H_2O) exopeptidase (R = large hydrophobic side chain).

The enzyme is a mammalian digestive enzyme, synthesized in the pancreas as inactive procarboxypeptidase, and trypsin activated in the small intestine.

Table 7.2 Selected lyases and isomerases

- *Carbonic anhydrase* (human): Zn^{2+}, early crystal structure determination (see text).

$$CO_2 + H_2O \rightleftharpoons HCO_3^- + H^+$$

- *Aconitase* (bovine heart): One of the irons in the [4Fe–4S] cluster is the site of substrate coordination and activation. Unusual nonredox role for an FeS cluster (Section 6.2).

$$
\underset{\text{citrate}}{
\begin{array}{l}
H_2-C-CO_2^- \\
HO-C-CO_2^- \\
H_2-C-CO_2^-
\end{array}}
\rightleftharpoons
\underset{\textit{cis}–\text{aconitate}}{
\begin{array}{l}
H-C-CO_2^- \\
\ \ \ C-CO_2^- \\
H_2-C-CO_2^-
\end{array}}
\rightleftharpoons
\underset{2R,3S\text{-isocitrate}}{
\begin{array}{l}
HO-CH-CO_2^- \\
H-C-CO_2^- \\
H_2-C-CO_2^-
\end{array}}
$$

- *D-xylose isomerase* (Arthrobacter): requires two divalent metal ions (Mn, Co or Mg) for catalytic activity. Also converts, more slowly, D-glucose to D-fructose.

$$
\underset{\text{D-xylose}}{OHC-C-C-C-CH_2OH}
\longrightarrow
\underset{\text{D-xylulose}}{HOH_2C-C-C-C-CH_2OH}
$$

Used industrially to produce fructose-enriched corn syrup (sweeteners). See Section 4.4(a).

7.3 Carbonic anhydrase

One of the most studied metalloenzymes, and one whose mechanism of action is well understood, is carbonic anhydrase. This name will be found in most of the literature although the correct one, carbonate dehydratase, is now being used occasionally. This enzyme occurs in nearly all phyla and has a MW ~ 30 kDa. Most studies (including a crystal structure) have been with the high activity human carbonic anhydrase isoenzyme, HCA II. It catalyses the reactions:

$$CO_2 + H_2O \rightleftharpoons H_2CO_3 \rightleftharpoons H^+ + HCO_3^- \qquad (7.1)$$

In erythrocytes, in which the enzyme is abundant, the proton generated in eqn 7.1 attaches to oxyhemoglobin thereby releasing oxygen (Section 8.7). HCO_3^- interchanges with extracellular Cl^-. Cystic fibrosis results from genetically caused misregulation of the chloride channels used for this interchange.

Hydration of CO_2 is quite slow ($t_{1/2} \sim 20$ s) at neutral pH and 25 °C. This half-life is dramatically reduced to microseconds by the enzyme in a solution saturated with CO_2. Zinc ion lies at the base of a deep pocket about 15 Å from the protein surface. The Zn^{2+} is coordinated to three histidines (94, 96 and 119) and one or two waters (low pH) or one hydroxyl group (high pH) (Fig. 2.5). Coordination by three protein-provided ligands and a water is a common feature of catalytic zinc in metalloenzymes. On the basis of a large amount of structural, kinetic and chemical data, the mechanism shown in

Fig. 7.1 is favoured for catalysis by carbonic anhydrase. It has many of the basic features exhibited by any metalloenzyme, although again there will be differences in detail.

It should be emphasized that no mechanism can be proved. The proposed mechanism must be consistent with the rate data if these are available. Enough other information on the reaction (detection of intermediates, determination of bond cleavage, isotopic distribution etc.) may then be amassed such that one can be fairly certain of the validity of the mechanism.

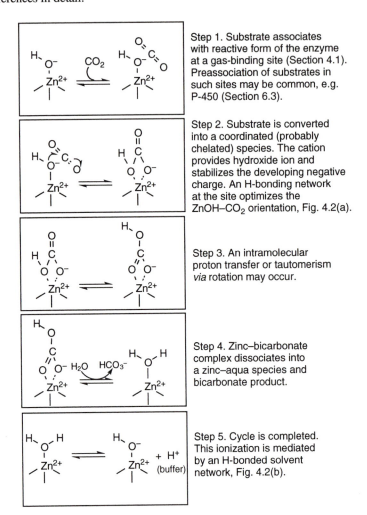

Step 1. Substrate associates with reactive form of the enzyme at a gas-binding site (Section 4.1). Preassociation of substrates in such sites may be common, e.g. P-450 (Section 6.3).

Step 2. Substrate is converted into a coordinated (probably chelated) species. The cation provides hydroxide ion and stabilizes the developing negative charge. An H-bonding network at the site optimizes the $ZnOH$–CO_2 orientation, Fig. 4.2(a).

Step 3. An intramolecular proton transfer or tautomerism *via* rotation may occur.

Step 4. Zinc–bicarbonate complex dissociates into a zinc–aqua species and bicarbonate product.

Step 5. Cycle is completed. This ionization is mediated by an H-bonded solvent network, Fig. 4.2(b).

Fig. 7.1 Steps in the carbonic anhydrase catalysed hydration of CO_2.

7.4   Cobalamin-dependent reactions

There are about a dozen enzymes which depend on the presence of cobalamins for their activity. These are mainly lyases and mutases, justifying the inclusion of cobalamins in this section, even though the coenzymes themselves are involved in redox reactions. There are two derivatives which concern us. Coenzyme B_{12}, which has been described as Nature's most beautiful cofactor (Fig. 7.2), assists in a 1,2 shift, which is normally difficult to achieve in organic chemistry ($X = NH_2$, OH or CR_3):

$$-\overset{\overset{\textstyle H}{|}}{\underset{\underset{\textstyle |}{|}}{C}}-\overset{\overset{\textstyle X}{|}}{\underset{\underset{\textstyle |}{|}}{C}}- \;\rightleftharpoons\; -\overset{\overset{\textstyle X}{|}}{\underset{\underset{\textstyle |}{|}}{C}}-\overset{\overset{\textstyle H}{|}}{\underset{\underset{\textstyle |}{|}}{C}}- \tag{7.2}$$

Thus a combination of D-α-lysine mutase and coenzyme B_{12} catalyses the conversion of a primary to a secondary amine:

$$\tag{7.3}$$

These reactions probably proceed via homolytic cleavage of the Co–C bond (which is relatively weak), to give the adenosyl radical ($\bullet CH_2R$) which mediates the 1,2 shift:

$$Co^{III}-CH_2R \;\rightleftharpoons\; Co^{II} + {}^\bullet CH_2R \tag{7.4}$$

$$-\overset{\overset{\textstyle H}{|}}{\underset{\underset{\textstyle |}{|}}{C}}-\overset{\overset{\textstyle X}{|}}{\underset{\underset{\textstyle |}{|}}{C}}- + {}^\bullet CH_2R \;\longrightarrow\; -\overset{\overset{\textstyle \bullet}{}}{\underset{\underset{\textstyle |}{|}}{C}}-\overset{\overset{\textstyle X}{|}}{\underset{\underset{\textstyle |}{|}}{C}}- + CH_3R \tag{7.5}$$

$$-\overset{\bullet}{\underset{|}{C}}-\overset{X}{\underset{|}{C}}- \;\rightleftharpoons\; -\overset{X}{\underset{|}{C}}-\overset{\bullet}{\underset{|}{C}}- \tag{7.6}$$

$$CH_3R + -\overset{X}{\underset{|}{C}}-\overset{\bullet}{\underset{|}{C}}- \;\longrightarrow\; {}^\bullet CH_2R + -\overset{X}{\underset{|}{C}}-\overset{H}{\underset{|}{C}}- \tag{7.7}$$

$$Co^{II} + {}^\bullet CH_2R \;\rightleftharpoons\; Co^{III}-CH_2R \tag{7.8}$$

The other cobalamin derivative of interest here is methylcobalamin, Fig. 7.2. It is one of the three cofactors with methylating abilities used by the enzyme system methionine synthase, which catalyses the reaction in eqn 7.9.

It has been suggested that on transfer of the CH_3 group to homocysteine, a four-coordinate cobalamin (eqn 7.10), containing cobalt(I), may be an intermediate ($CH_3THF_0 = N$,5-methyltetrahydrofolate, the second cofactor).

$$\begin{array}{c} {}^+NH_3 \\ | \\ CH(CH_2)_2SH \\ | \\ CO_2^- \end{array} \quad\xrightarrow[-\,H]{+\,CH_3}\quad \begin{array}{c} {}^+NH_3 \\ | \\ CH(CH_2)_2SCH_3 \\ | \\ CO_2^- \end{array} \tag{7.9}$$

homocysteine   methionine

The recent determination of the crystal structure of the enzyme system from *E.coli* has shown how methylcobalamin binds to methionine synthase and has provided some understanding of the role of the coenzyme in the methylation reaction. A histidine from the protein replaces the dimethylbenzimidazole in methylcobalamin, when it is bound to the protein.

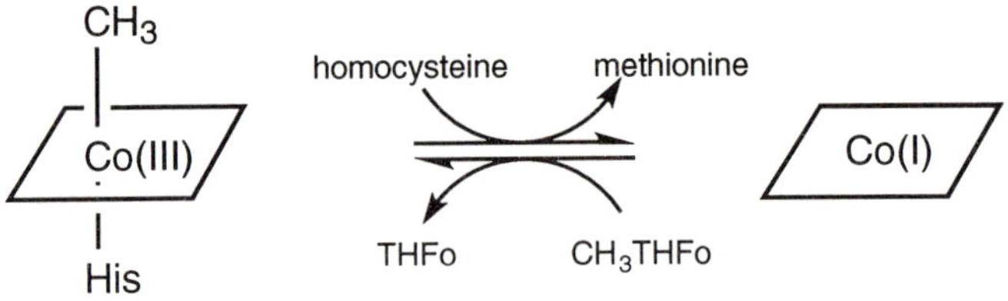

$$CH_3 - Co(III) - His \quad \overset{\text{homocysteine} \quad \text{methionine}}{\underset{\text{THFo} \quad CH_3THFo}{\rightleftharpoons}} \quad Co(I) \quad (7.10)$$

All of the genes required for vitamin B_{12} biosynthesis have been cloned, sequenced and expressed. The many steps, each using a specific enzyme, have now been mapped out. Remarkably, microorganisms achieve the synthesis of vitamin B_{12} *in vivo* with complete control of the stereochemistry at the nine chiral centres in the corrin ring. In *Psuedomonas denitrificans* (the commercial source of the vitamin) the cobalt is inserted at a quite late stage in the biosynthesis.

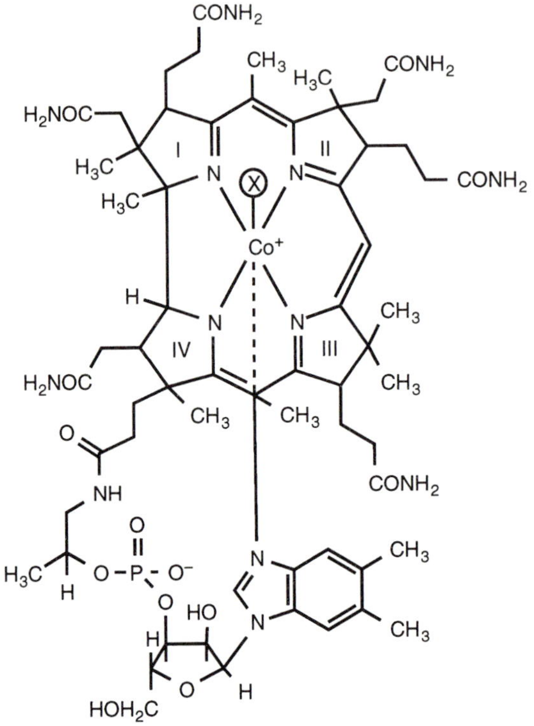

Fig. 7.2 The structure of cobalamin derivatives. The corrin ring system is slightly bent. In vitamin B_{12}, X = CN^-; methylcobalamin, X = CH_3 and coenzyme B_{12}, X = 5'-deoxyadenosyl. The metal–C bond is very rare in biochemistry.

7.5 Transport and storage of d-block elements

The transition metal ions must be moved from their point of ingestion to the organs which require them. As with the s-block cations, there is an unequal cellular distribution of transition metal ions (e.g. Cu is mainly extracellular while Zn is mainly intracellular) so that cation transport across biological frontiers is necessary. There are, however, big differences between the d- and s-block cations. The transition metal ions precipitate at biological pHs and are often toxic because they can produce small amounts of radicals by reactions such as (Section 8.6):

$$Fe(II) \; + \; O_2 \; \rightarrow \; Fe(III) \; + \; O_2^- \qquad (7.11)$$

$$Fe(II) + H_2O_2 \rightarrow Fe(III) + OH + OH^- \qquad (7.12)$$

They thus require carrier ligands for transport. Although some detail is lacking, iron is the best understood of the transition metals, both because of its importance and of the relative availability of material (e.g. in humans). The modes of transport of the d-block elements differ in bacteria and vertebrates.

Iron in bacteria

Nearly all microorganisms require iron for growth. In bacteria, iron is usually solubilized and transported by non-protein chelating ligands, called siderophores. These are widespread and have been found in maritime microorganisms and in plants (phytosiderophores, Gk: plant iron carriers). Siderophores provide O-donors and chelate Fe(III) very strongly with binding constants greater than 10^{30} M^{-1}. The usual chelating centres are hydroxamate, o-dihydroxy aromatic or α-hydroxycarboxylate groups. All these types of groups are present in pseudobactin (Fig. 7.3).

Q. Suggest the coordination sites in the phytosiderophore mugineic acid, which is produced by plants for iron metabolism. How many chiral centres are present?

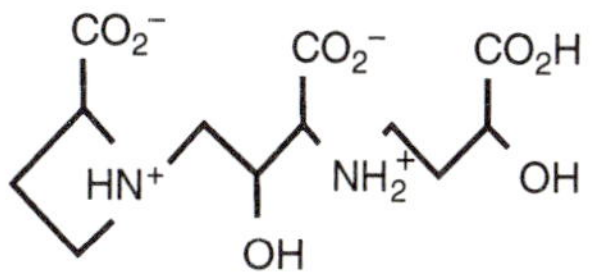

Certain rhizobacteria produce extracellular siderophores such as pseudobactin, which sequester iron and make it unavailable to disease causing microorganisms. Their growth is thereby inhibited, and so that of plants they infect (e.g. potatoes, carrots) is promoted as their diseases are controlled.

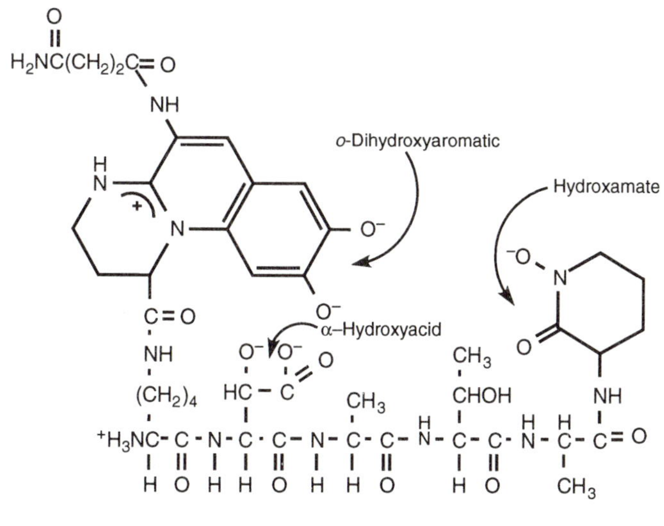

Fig. 7.3 The structure of pseudobactin. This siderophore is a linear hexapeptide with chelating side groups so disposed as to form an O_6 octahedron around Fe(III).

Three hydroxamate groups are contained in the clinically important desferrioxamine B, which is used for the treatment of iron overload diseases (Section 7.8). The siderophores are too large to permeate the water filled channels of the outer membrane so they are recognized there by receptor proteins and thereby are moved across the inner membrane to the cytoplasm by an ATP driven reaction (Fig. 7.4). A vexing problem for some time has been understanding how the iron is released from the ligand. One reasonable idea is that the Fe(III) complex is reduced enzymatically to the Fe(II) form which is known to dissociate (to Fe^{2+} + free ligand) much more easily ($K \sim 10$ M^{-1}) than does the oxidized form.

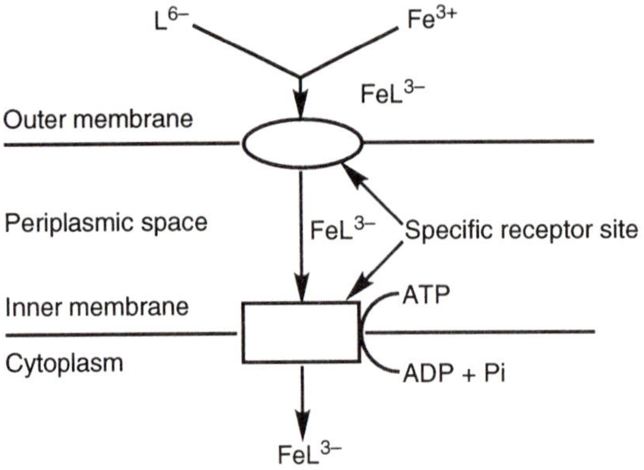

Fig. 7.4 Model for ATP driven, enterobactin (L^{6-}) mediated, Fe uptake in *E. coli*. Ferric enterobactin (FeL^{3-}) interacts with specific receptors on the outer and inner membranes of *E. coli*. Variations of this model have been proposed for other siderophores.

Iron in vertebrates

This discussion is concerned mainly with the transport and storage of iron in humans. In vertebrate species, iron is efficiently recycled and little is absorbed from the diet or is excreted. Only about 1% of the body's iron is being used at any time. The remainder is stored. Proteins are used here for the storage and transport of iron and by this means, the sequestering of Fe prevents its easy access and use by parasitic microorganisms.

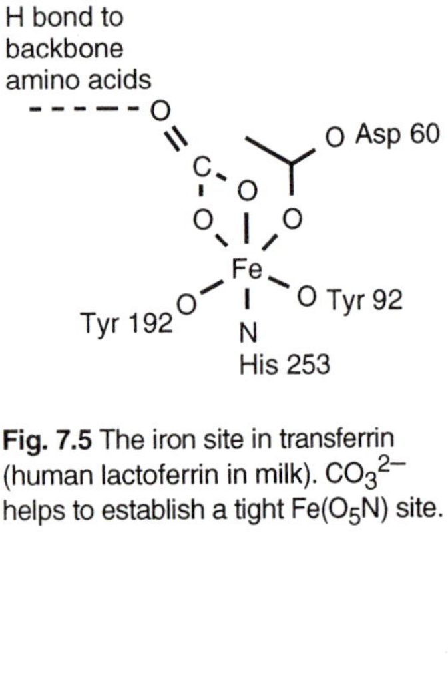

Transferrin

Iron is transported from the stomach to the various sites for metabolic action by transferrin. It is a glycoprotein, MW~80 kDa, consisting of two homologous lobes. Each lobe is made up of two domains which contain a single high affinity metal binding site in the interdomain cleft. One iron is present in each lobe and the two irons are similar structurally (Fig. 7.5) but behave differently. The irons are buried about 10 Å below the protein surface and are strongly bound ($K \sim 10^{20}$ M^{-1}). There is an important hinge motion in the protein associated with the addition of Fe(III) and CO_3^{2-} to apolactoferrin. This results in the transferrin receptor proteins (which are membrane-bound) recognizing iron containing lactoferrin but not the apo form. As with the siderophores, it is uncertain how iron is released, but lowered pH and CO_3^{2-} may play a role. Once in the cell, the iron is used quickly for enzyme and heme manufacture or for storage by ferritin.

Fig. 7.5 The iron site in transferrin (human lactoferrin in milk). CO_3^{2-} helps to establish a tight $Fe(O_5N)$ site.

Ferritin

Although iron has been detected in the brain as magnetite, Fe_3O_4, it is stored in animals (also in plants and some bacteria) mainly in a non-toxic form in a remarkable piece of biological machinery called ferritin. In humans, it is especially high in concentration (~0.1 M) in liver, spleen and bone marrow. Ferritins are a family of large, approximately spherical (diameter ~130 Å), proteins enclosing a mineral core of ~75 Å in diameter, most of which is not directly bound to the protein. As many as 4500 irons (but usually less) are present in the core as hydrated ferric oxide, $Fe_2O_3.nH_2O$, associated with a variable phosphate content. The approximate core composition is $[FeO(OH)]_8[FeO(PO_4H_2)]$. The protein portion of ferritin consists of 24 polypeptide subunits, each coiled as in a capsule. Hydrophilic- and hydrophobic-lined channels connect the inner core to the outer environment and allow iron and organic reductants to enter and leave.

Other transition metal ions

The trace amounts of most transition metals present in the body makes the study of the mechanisms of their transport and storage extremely difficult. Metallothioneins are used extensively in the animal and plant kingdoms. They are rich in cysteine and therefore have a preference for soft metal ions such as Cu^{2+} and Zn^{2+}, for which they act *in vivo* as homeostatic control agents. The production of metallothioneins is induced by a number of toxic transition metals (Cd, Hg, Ag, Au) which activate the transcription of specific genes. These proteins thus also act as detoxifying agents. The structures of metallothioneins have been thoroughly examined both in solution (NMR) and as solids (X-ray crystallography, EXAFS), Fig. 7.6. The value of metallothioneins as metal ion carriers is consistent with the observations that they interchange metal readily within the cluster, and easily exchange their metal ions with metal ions in other clusters and with free metal ions in solution.

Livers of new-born rats contain about 20 times the amounts of Cu and Zn metallothioneins as those of 70 day old rats. These metal ions are therefore stored as metallothioneins and used after birth.

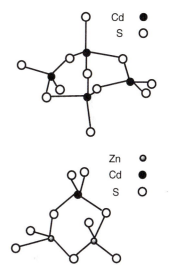

Fig. 7.6 Metal clusters in rat liver metallothionein. The protein isolated from rat liver has 61 amino acids 20 of which are cysteines, and 5Cd and 2Zn per molecule. There are two clusters; a Cd_4S_{11} adamantane like and a $CdZn_2S_9$ chair structure. It is the only known protein which contains Cd in the native state.

7.6 The d-block metals, DNA and RNA

The importance of the s-block Mg^{2+} ion in nucleic acid chemistry is clear (Section 5.4). It is however both in simple complexes and in metalloproteins where the transition metals have important roles to play in their interactions with nucleic acids.

Metal complexes

The binding of a Pt(II) complex to DNA is an important first step in its anticancer effect (Section 7.8). This finding, in part, promoted a general interest in the interactions of metal complexes with DNA. The metal centre in these complexes can bind directly to DNA (Fig. 2.4). Complexes containing all chelated ligands are more likely to retain them and to form non-coordinate complexes with nucleic acids. If the chelates are planar heterocyclic ligands, they will often intercalate, usually in a major groove; that is they stack in between DNA base pairs (Fig. 2.9) and in the process distort the DNA helix. Most complexes, and small molecules in general, will associate with DNA in the minor groove. However, they can be made to intercalate by attachment of the metal complex to an intercalator, an example of which is the specifically designed complex MPE–Fe(II) (Fig. 7.7). Reaction of this complex with H_2O_2 and a reducing agent generates hydroxyl radicals. Since the methidium segment (the aromatic portion of the molecule) intercalates randomly in DNA, the sugar segments on the DNA will be oxidized randomly by the hydroxy radicals. MPE–Fe(II) is therefore a footprinting agent for large and small molecules.

Footprinting techniques determine the region of DNA which is bound to small (e.g. drugs) or large (e.g. proteins) molecules. The DNA adduct is treated with a footprint reagent. The products are denatured and separated on a gel. The DNA adduct shows up as a blank spot or footprint in the position on the gel of DNA protected by the bound material.

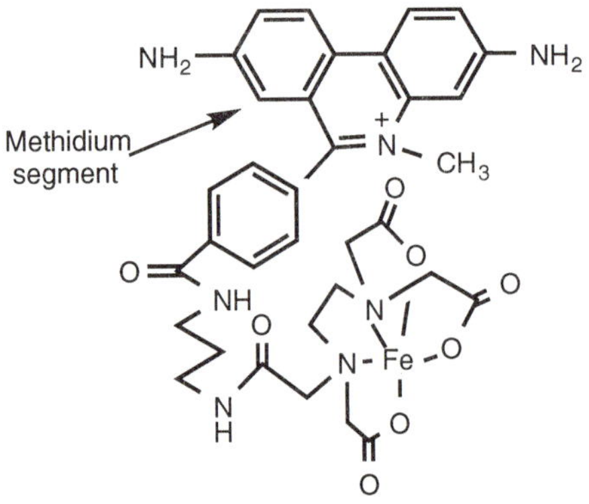

Methidium segment

Fig. 7.7 Structure of the intercalating agent MPE–Fe(II).

A deficiency of zinc in the diet leads to delayed development. This arises from the inability of estrogen receptors to fold properly without zinc. In these fingers, zinc is coordinated tetrahedrally to four cys.

Metalloproteins

In their interactions with nucleic acids, the metal ion or metal cluster portion of a protein may have a catalytic, structural or regulatory role. These metalloproteins can function in both repair and transcription processes. Endonuclease III is a DNA repair enzyme that removes oxidized pyrimidines from DNA and leaves a single stranded nick at that point. The enzyme contains an Fe_4S_4 cluster which is distant from the DNA binding site in the protein. The cluster is very stable toward oxidation and reduction (which is very unusual, Table 6.2) and probably acts only to position certain basic residues in the enzyme so that they can interact with the DNA. In this respect, the Fe_4S_4 cluster plays a similar role to that of Zn^{2+} in zinc fingers. An example of the latter is the transcription factor IIIA (Fig. 4.6). Nine fingers exist in IIIA and each finger binds to appropriate regions of DNA and RNA. Since the discovery of the first zinc finger in the mid 1980s, very many more such eukaryotic regulatory proteins, with different zinc binding patterns, have been found in many species. They contain any number from 2 to 40 tandem fingers which can interact with DNA in a number of ways (usually at a major groove). This variety allows the cell to cleverly produce a vast collection of different transcription factors by changing the order and number of modules ('pick and mix').

Q. The melting temperature of calf thymus DNA increases on addition of 1 mole Mg^{2+} / mole DNA but decreases when treated with 1 mole Cu^{2+} / mole DNA. What is a reasonable explanation for this behaviour?

Q. Why do you think that zinc has been singled out for use in Nature in nucleic acid binding proteins?

Metalloregulatory proteins may also act as triggers, repressing or activating transcription depending on the concentration of the metal ion. The MerR system regulates mercury resistance in bacteria, which depends on the expression of Hg detoxification genes. In the absence of Hg(II), MerR binds tightly to the promoter and thus inhibits its binding by RNA polymerases. Hg(II), even in low concentration (nM), binds specifically and tightly to DNA-bound MerR, promotes a conformational change (some unwinding) which now facilitates the binding of RNA polymerase and so induces the expression of the gene family and the production of detoxifying enzymes. A number of systems involving gene expression are now known to respond to metal ions in prokaryotes and eukaryotes. The specificity to a metal ion requires that an unusual geometry, specific for that metal ion, must exist in the metal regulatory protein. In the case of MerR this appears to be a 3 cys site in a trigonal plane, which does not easily accommodate other soft metal ions such as Zn(II) and Ag(I).

7.7 Biorecovery of metals from the soil

Mining

We have seen that microorganisms are used in the deposition of minerals for use by living organisms (Table 5.4). Conversely, acidophilic bacteria, such as *Thiobacillus ferrooxidans* can be used at temperatures ranging from 30 to 70 °C to leach minerals from ores during growth. Such bacteria, which generate acid (a pH as low as 1 may be attained), have been termed chemolithotrophic ('rock eating'). They aid in the breakdown and solubilization of insoluble mineral sulfides by oxidation, either directly with O_2:

$$CuS + 2O_2 \rightarrow CuSO_4 \qquad (7.13)$$

or indirectly via Fe(III) produced from Fe(II) by bacteria.

Such processes are likely to become more important as higher grade ores become in short supply and the microbiological leaching of low grade ores becomes economically viable. Even now approximately one quarter of the world's production of copper is by microbially supported leaching. The use of microorganisms to recover toxic metals from contaminated dumps is also likely to be increasingly exploited.

Toxic metal cleanup

Some plants are capable of concentrating metals from the soil to a level of several percent of their dry weight. They accumulate amounts which would kill most plants, probably as a defense mechanism against plant-eating insects. A number of such plants (hyperacumulating) use phytochelatins (cysteine-glutathione small peptides), organic acids or amino acids such as histidine to form complexes with these metal ions. It is hoped that these plants might be used to clean up contaminated sites (phytoremediation) such as in abandoned mines (Zn, Pb) and municipal waste sites (Cu, Hg, Pb).

7.8 Health and the d-block elements

The use of transition elements, both in the metallic form or in compounds, as medicines (Table 7.3) is ancient. Copper, gold and zinc were all used in Arabic, Hindu and Chinese prescriptions, although even now their mode of action is not well understood. Sometimes the metal ion is the important component of the pharmaceutical (Gd and Au). In some cases the use of the compound *in toto* is vital (Tc, Pt and Co). Table 7.3 shows some important inorganic pharmaceuticals as well as diagnostic and therapeutic agents for serious abnormalities.

Acidophilic bacteria are autotrophic bacteria growing in the absence of organic material and fixing CO_2 from the atmosphere.

Heavy metal (e.g. Cd, Cu) deposits and tree deaths appear to be related. Phytochelatins can sequester these metal ions and their concentrations in the tree species in decline (e.g. red spruce) increases as forest damage increases. This however, leads to glutathione deficiency and less protection against oxidative damage to the tree.

Early trials using alpine pennycress (*Thlaspi caerulescens*) and mustard plant (*Brassica juncea*) have shown promise for toxic metal cleanup. It is hoped to identify the gene responsible for the hyperacumulating property for transfer to plants which are better growing, larger and non-edible (to small mammals). The harvested plant could be burned to recover valuable metals (Cu, Ni).

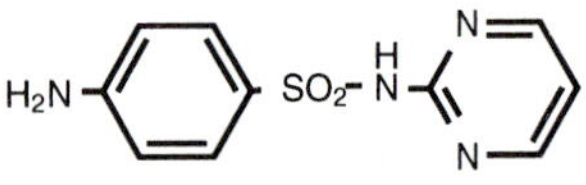

Sulfadiazine

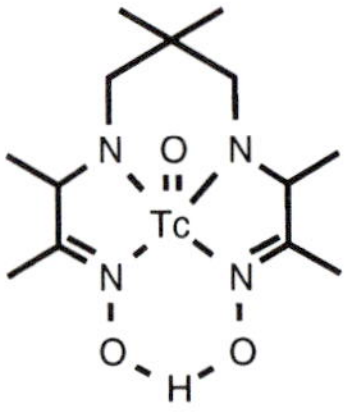

Tc(HMPAO)

$^-O_2CCH_2N[(CH_2)_2N(CH_2CO_2^-)_2]_2$

DTPA

Table 7.3 Pharmaceuticals derived from the transition elements

Compound	One trade name	Value and notes
Zr(IV) glycinate		Antiperspirant
Vitamin B12	Ce - cobalin	Food supplement
Ag(I) sulfadiazine	Flamazine	Antibacterial for severe burns
$ZnSO_4 \cdot H_2O$	Z–span	Food supplement
Zn oxide or carbonate (trace of Fe_2O_3)	Calamine lotion	Antimicrobial and antifungal in skin ointments
$Tc(CNR)_6^+$ [R = $CH_2C(CH_3)_2OMe$]	Cardiolite	Used to image heart abnormalities
Tc(HMPAO)	Ceretec	Cerebral perfusion imaging
$Gd(DTPA)^{2-}$	Magnevist	Improves magnetic resonance imaging scans, administered in up to 10 g doses
cis-$Pt(NH_3)_2Cl_2$	Cisplatin	Cytotoxic drug, effective in treatment of cancers of testes or ovaries
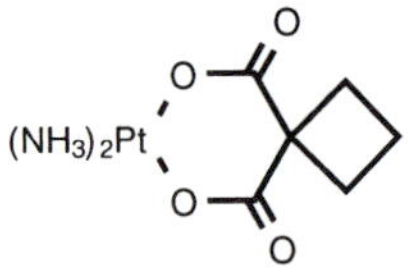	Carboplatin	Second generation less toxic cytotoxic platinum drug
cis-$(CH_3COCHCOC_6H_5)$- $Ti(OEt)_2$	Budotitane	On trial for treatment for colon cancer, Ti delivers effective ligand
$Au(CH_2(CO_2^-)CH(CO_2^-)S)$	Myocrisin	Antiarthritic
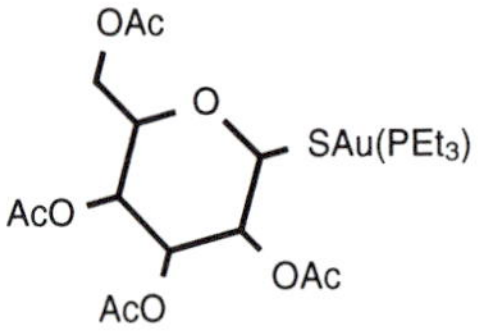	Auranofin	Oral agent for rheumatoid arthritis

Clinical use of ligands

Ligands, and particularly organic chelating agents, have a very special clinical role in handling the transition metal ions in the body and in facilitating their external administration. Chelating ligands have been used to:

(a) remove damaging metal ions

For this purpose a ligand which can bind strongly to the metal ion in question is required. In addition, it should be quite specific for that metal so that other beneficial metal ions are not removed. The resulting complex should be non-toxic and water soluble so that it can be easily excreted. Excess, and potentially lethal, amounts of metal ions can accumulate in the body for various reasons. Copper accumulation in the liver and the brain arising from a genetic misfunction in the copper storage system in the body (Wilson's disease) can be up to 10–15 grams instead of the normal 100–150 mg. This constitutes a potential disaster (dementia and eventual death) which can be controlled by the administration of (for life, unfortunately) approximately one gram daily of D-penicillamine (the L- form is toxic). Yet another genetic disorder (thalassemia) can lead to the body being unable to synthesize one or more globin chains correctly. Blood transfusions are necessary to correct the deficiency, but this treatment itself leads to the accumulation of toxic amounts of iron in the body, rising to as much as 70 grams in 10 years where normally it is 4–5 grams. Chelation therapy, using the siderophore desferrioxamine B (Desferal), Fig. 7.8, can reduce the iron concentration to reasonable amounts, but painful multiintravenous injections are necessary.

Gold has been used to treat leprosy, rheumatic fever, tuberculosis and even mortality. Au(I) compounds are used as antirheumatic agents but side effects can be severe. The mechanisms of action are unclear.

Thalassemia: Derived from *thalassa* (Gk. sea) since originally thallassaemia was diagnosed in children from the Mediterranean area. It is now known to be widely distributed. A 26 year old Chinese sufferer who had over 400 blood transfusions is reported to have set off the metal detector at an airport security checkpoint because of his high iron content. Without treatment, few patients reach the age of ten. With a combination of transfusions and administration of desferrioxamine, patients are still surviving in the 25 years or so since this clinical procedure was adopted.

(CH$_3$)$_2$C —— CHCO$_2$H
    | |
   HS   NH$_2$

D-penicillamine

HS — C — CO$_2$H (with H above)
      |
HS — C — CO$_2$H (with H below)

2,3-dimercaptosuccinic acid

N with CH$_2$CO$_2^-$ and CH$_2$CO$_2^-$
|
CH$_2$
|
CH$_2$
|
N with CH$_2$CO$_2^-$ and CH$_2$CO$_2^-$

EDTA

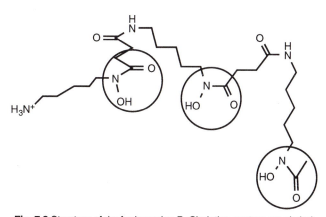

Fig. 7.8 Structure of desferrioxamine B. Chelation centres are circled.

Searches for other iron chelating agents which will avoid this problem are currently underway. The treatments of the defects associated with Cu and Fe imbalances will almost certainly be aided by the discovery of the genes responsible for these diseases.

Poisoning of human beings can occur on ingestion of certain metal ions or compounds (Table 1.6), for example in an industrial accident. The ligands D-penicillamine, Desferal, 2,3-dimercaptosuccinic acid and EDTA have all been effective as treatments for the removal of the heavy metal ions Hg(II), Pb(II), Cd(II) and radioactive [239]Pu from most human organs.

Q. EDTA and derivatives have major roles in chelation therapy. Care must be taken however in their use. Hypocalcemia, resulting from a rapid drop in [Ca^{2+}] in blood plasma can cause cramps, convulsions and even death. How would you control this?

(b) target metalloenzymes

Metalloenzymes, as well as being vital biological catalysts, can also promote undesirable reactions. Inhibitors, in the form of ligands, may prevent or control this unwanted function by binding to the metal ion. Three examples, all directed at zinc enzymes, illustrate this use of ligands (Table 7.4).

Collagenases are part of the important family of matrix metalloproteinases (MMP) which are zinc enzymes responsible for catalysis of the degradation of connective tissues such as tendons and skin in vertebrates. They are also responsible, incidentally, for resorption of the tadpole tail connective tissue during metamorphosis. Inhibitors of these enzymes are being used in clinical trials for treating diseases in which MMPs are implicated such as rheumatoid and osteoarthritis (cartilage in joints is degraded by an MMP) and for cancer ('cancer pills') since the observation that MMPs are overproduced in cancers.

Table 7.4 Control of undesirable effects of key metalloenzymes

Metalloenzyme	Reaction catalysed	Inhibitor
Angiotensin-converting enzyme (ACE)	Phe–his bond cleavage in decapeptide (angiotensin I) → octapeptide (angiotensin II) which is a potent hypertensive (blood-pressure raising) agent.	EDTA or captopril bind to Zn and control hypertension (vasodilators). captopril
Enkephalinases	Gly–phe bond cleavage in neuropeptides (enkephalins) → inactive fragments. Pain level control lost.	Thiorphan binds to Zn and acts as an analgesic. thiorphan
Collagenases	Digest matrix framework (although important in bone remodelling and wound healing their activation from zymogen forms must be carefully regulated).	Tissue inhibitors are small proteins which interact with Zn by ligation.

(c) transport metal as a complex to sites for organ targeting

This approach is used for diagnosis and treatment of organ malfunctions, usually cancers, and the complex *in toto* must be transferred to the site. We will cite one very effective example for each category.

Diagnosis: Technetium complexes fulfil the requirements of an ideal diagnostic agent to such a large degree, that about 90% of clinical diagnostic imaging procedures involve Tc. ^{99m}Tc is an easily detected powerful γ-emitter. It is short lived ($t_{1/2} = 6$ h) and the emitted photons can penetrate, but not damage, tissue. Technetium, being in the Mn, Tc, Re triad (s^2d^5) is able to form a variety of metal complexes in different oxidation states (+1 to +7) which are kinetically stable *in vivo*. Happily, it has been observed that different Tc complexes become localized in specific tissues and cells, particularly in abnormalities, although the reasons for this are not clear. Examples are shown in Table 7.3. The radioactive Tc complex is injected into the patient. After about three hours the Tc clears from the blood and soft tissues, accumulates in the bone and is there camera scanned.

Therapy: Behind the present day use of *cis*-Pt(NH$_3$)$_2$Cl$_2$ (cisplatin), Table 7.3, as an effective agent for the treatment of certain types of cancers is a fascinating story of an accidental discovery of an unusual phenomenon, tenacious follow up of the implications and finally the production of a drug of great clinical value. A great deal of research has now shown that an effective cytostatic platinum complex must have a square planar disposition of the four ligands around Pt(II), with two *cis*-primary or secondary amine ligands and two (relatively easily hydrolysed) chloride or carboxylate groups. The fundamental effectiveness of cisplatin stems from its ability to bind to DNA (Fig. 2.4) and block replication. The major product is Pt(II) bound to adjacent N7 guanines on the same strand to form a chelate complex (intrastrand crosslink, Fig. 7.9). This binding disrupts base–base stacking interactions in DNA and finally leads to a kink in the B-DNA helix, which induces binding of specific cellular proteins (other platinum complexes cannot do this). The precise mechanism after this stage is uncertain and complex. One possibility is that the platinum-invoked lesions are repaired faster in non-tumor tissue, leading to the death of the cancer cells.

Since the first use of cisplatin, hundreds of platinum complexes have been tested for antitumor activity. Carboplatin (Table 7.3) is now used and platinum(II) complexes such as (NH$_3$)$_2$ClPtNH$_2$(CH$_2$)$_n$H$_2$N–PtCl(NH$_3$)$_2^{2+}$ are showing, in preclinical tests, activity towards cisplatin resistant tumors. This may be related to differences in their DNA-binding characteristics (interstrand crosslinks and nonbending). Proton and ^{15}N NMR have been very useful for investigating the speciation and reactivity of the Pt(II) drugs in *intact* biological fluids. Concentrations of Pt species as low as 20 μM can be observed in the milieu of much larger concentrations of numerous other species.

Current approaches: One ploy to broaden the scope of diagnostic and therapeutic methods is to attach the appropriate metal ion to an entity which tends to locate in specific parts of the body. Phosphonates, containing the –CH$_2$PO$_3$R$^-$ group, have a high affinity for growing bone. Therefore the ^{99m}Tc methylenediphosphonate complex is used for diagnosing bone malformities and Pt phosphonate complexes show promise for treating bone tumours. The radioactive complex ^{153}Sm(X$_2$N(CH$_2$)$_2$NX$_2$)$^{5-}$, where X=CH$_2$PO$_3^{2-}$, appears to be useful in alleviating the pain caused by bone tumours.

Monoclonal antibodies are being used as carriers for radionuclides in order to target antigen sites associated with cancer and other diseases. Although results are mixed, some success has been reported. The radiometal chelate complex is attached directly to the antibody, but sufficiently distant so that it does not interfere with the immunoreactivity of the antibody.

An electric field was found to inhibit bacterial cell division, but not growth. This effect was traced to a platinum complex, formed from the partial dissolution of the Pt from the electrodes into the medium. This very surprising result led to the discovery that the Pt(II) complex was a very effective antitumour agent. Serendipity is a common occurrence in science. Genius lies in the appreciation and exploitation of a lucky discovery. For a full account consult Further Reading.

Q. Why do you think that *cis*-Pd(NH$_3$)$_2$Cl$_2$ and *trans*-Pt(NH$_3$)$_2$Cl$_2$ are ineffective cytotoxic agents?

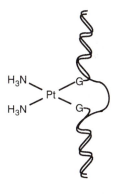

Fig. 7.9 Intrastrand crosslink in cisplatin. The DNA is unwound and bent toward the major groove (Fig. 2.9). The repair of this lesion features in the mechanism of action of, and in the acquired resistance to, the drug.

8 The p-block

8.1 Introduction

In this block of elements, filling of the three p-orbitals is being completed, using six electrons in each series, and giving rise to Groups 13–18. The lighter members, which are of most importance in biology, are non-metals. The elements become increasingly more electronegative as the p-orbitals fill, because there is a driving force to accept electrons to reach the very stable p^6 configuration of the noble gases. Elements such as oxygen, fluorine, chlorine and, to a certain extent nitrogen, are highly electronegative and are therefore able to take part in hydrogen bonds (Section 2.4). Some of the most important molecules in our environment are formed from elements of this group, i.e. O_2, N_2, CO_2, H_2, H_2O and others, and in the next sections we shall be concerned with their complex interrelationships. Strictly, the element hydrogen belongs to the s-block, but because the $2H^+/H_2$ couple is so important in biochemical cycles involving p-block elements, and because H has p-characteristics (i.e. it forms H^-) it is included here. The d-block elements (Chapters 6 and 7) are essential components of many of the metalloenzymes which catalyse bond cleavage and overcome the intrinsic low reactivity of N_2, H_2 and CH_4; control the extreme reactivity of O_2 and the toxicity of O_2^- and O_2^{2-}; and aid in the reduction of nitrogen oxide compounds. The cycles involved, and their exquisite interlinking, are illustrated in Fig. 8.1, which is obviously greatly oversimplified. Each cycle will now be examined.

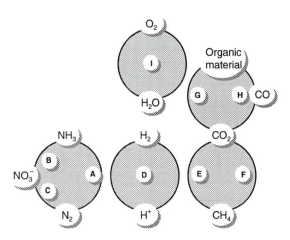

Fig. 8.1 The interlinking of the nitrogen, hydrogen, methane, carbon dioxide and oxygen cycles. These are the most important cycles in nature.

8.2 The nitrogen cycle, A–C

This cycle is relevant to animals, plants and bacteria. The transformations can occur in either direction but intermediates are well established, in some cases, in one direction only. A key reaction in the cycle is nitrogen fixation **A** (eqn 8.1) which is catalysed by nitrogenase:

$$N_2 + 8H^+ + 8e^- + 16MgATP \rightarrow 2NH_3 + H_2 + 16MgADP + 16P_i$$

$$(8.1)$$

The electrons are supplied by the Fe(II) form of ferredoxin and the ammonia produced is used for the synthesis of organic compounds required for cell growth. The enzymes which catalyse this reaction have been the focus of a great deal of attention for a number of years. Although there are forms which contain only vanadium or iron, the most common and most investigated enzyme contains both an Fe protein and a MoFe protein. Nitrogenase is found in certain soil bacteria and on the root nodules of legumes (e.g. pea and alfalfa). The crystal structures of the two proteins in nitrogenase from *Azotobacter vinelandii* have been solved, after a decade of active study. The Fe protein contains one Fe_4S_4 cluster which is very sensitive to oxygen. MgATP (and MgADP) binds about 20 Å from the FeS cluster, and its binding causes a substantial conformational change. The Fe protein transfers one electron, for each two molecules of MgATP hydrolysed, to the MoFe protein to which it is bound (probably) by salt bridges. This process is repeated until the MoFe protein has accumulated eight electrons which it then uses to effect the reduction of N_2 (eqn 8.1). The nature of the binding site for N_2 in the MoFe protein is unclear. It undoubtedly binds at one of the cofactors in the MoFe protein which has a fascinating structure (Fig. 8.2). The low coordination number (three) of some of the Fe sites *suggests* that one or more of these, and not Mo, as has been supposed, is the substrate binding site. Another cluster (cysteine-bridged 2[4Fe–4S]) is also located in the MoFe protein.

Rhizobium organisms in the legume root nodules 'fix' about 10^8 tons of nitrogen a year. This represents about 60% of all the nitrogen fixed, the remainder arising from fertilizers and lightning discharges (via the formation of nitrogen oxides, Fig. 8.3). The latter two processes produce nitrate ion, which is the preferred source of nitrogen for most green plants. It is converted into NH_3, **B** (eqn 8.2):

$$NO_3^- \xleftrightarrow{\ Mo\ } NO_2^- \xrightarrow[\text{siroheme}]{Fe_4S_4} NH_3 \quad \text{assimilation} \quad (8.2)$$

using *assimilatory* nitrate and nitrite reductases. When plants die the nitrogen-containing organic compounds are reoxidized to nitrate. However, because this cycle is far from 100% efficient, the shortfall must be made up by nitrogen fixation. Ammonia can be converted into NO_2^- and NO_3^-, **B** (eqn 8.3):

Leghemoglobin is also contained in the nitrogen-fixing nodules of legumes. It is monomeric and has a high affinity for oxygen, thus reducing the free concentration of oxygen, which is very toxic, in the nitrogen-fixing region of the plant.

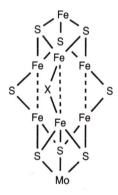

Fig. 8.2 Structure of $MoFe_7S_8$ in the MoFe cofactor of nitrogenase from *Azotobacter vinelandii*. The cofactor has [4Fe 3S] and [Mo 3Fe 3S] clusters bridged by 2S$^-$ ions and an as yet unidentified ligand X. Homocitrate is an essential part of the MoFe cofactor and is chelated to Mo (not shown). The tetrahedral ligation of the terminal Fe is completed by a protein cysteine.

Nitrification:

$$2NH_3 \xrightarrow{\text{Cu}} NH_2OH \xrightarrow{\text{heme d}} NO_2^- \rightleftharpoons NO_3^- \qquad (8.3)$$

Plants grow better in a NO_3^- medium than in one containing NH_4^+ (NH_3). The latter are toxic in higher concentrations, but are the sources of amino acids in syntheses. All nitrogen in the animal world originates from plants.

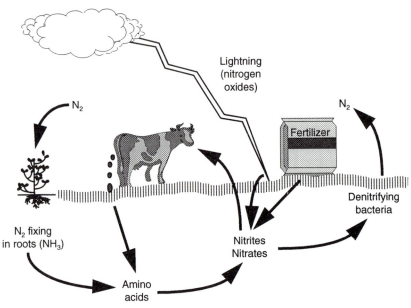

Fig. 8.3 The nitrogen cycle in nature. Mobile changes in the nitrogen-containing compounds circulating between the atmosphere, the soil and microorganisms.

Two of the enzymes involved in nitrification are ammonia monooxygenase and hydroxylamine oxidoreductase, but apart from the fact that they contain Cu and a d-type heme (Section 6.2) respectively, little is known yet about their structures or modes of action. Both heme- and copper-containing dissimilatory nitrite reductases have been described. These catalyse the conversion of NO_2^- to NO, which is one step in the anaerobic denitrification of NO_2^- or NO_3^- by prokaryotic organisms, **C** (eqn 8.4):

Denitrification: $\qquad\qquad\qquad\qquad\qquad\qquad\qquad\qquad (8.4)$

$$NO_3^- \underset{}{\overset{\text{Mo}}{\rightleftharpoons}} NO_2^- \xrightarrow{\text{Cu or heme } d_1} NO \xrightarrow{\text{heme}} N_2O \xrightarrow{\text{Cu}} N_2$$

This represents a loss of plant nutrients and is counter to N_2 fixation. Nitrate reductases always use Mo to which NO_3^- binds (Table 6.4). Unusual d-type hemes feature prominently in the steps described by equations 8.2–8.4. Nothing like the attention had been paid to the metalloenzymes which catalyse the steps in **A–C** as that afforded nitrogenase, but the former are being increasingly studied.

There are striking similarities in the UV–vis and resonance Raman spectral properties of a Cu site in nitrous oxide reductase, cytochrome c oxidase and dissimilatory nitrite reductase. The presence of tetrahedrally distorted type 1

Cu cysteinates in all of the proteins is indicated by the cys(S) → Cu(II) charge transfer band giving rise to a purple colour. The X-ray structures of nitrite reductase (*Achromobacter cycloclastes*), and cytochrome c oxidase (both *Paracoccus denitrificans* and bovine heart, Table 6.3) confirm this.

8.3   The proton–hydrogen interconversion, D

The hydrogen produced in the cytoplasm during the fixation of nitrogen (eqn 8.1) is trapped into relinquishing electrons, which can be used to fix more nitrogen and increase the efficiency of nitrogenase. This reaction is catalysed by the hydrogenases which are a variety of enzymes found in both aerobic and anaerobic bacteria. They catalyse the interconversion shown in eqn 8.5

$$H_2 \rightleftharpoons 2H^+ + 2e^- \qquad (8.5)$$

leading to consumption or production of hydrogen depending on the organism. There are a number of electron acceptors including O_2, NO_3^-, SO_4^{2-} and CO_2. For example, under anaerobic conditions, hydrogenases catalyse the reduction of CO_2 by H_2 to form CH_4 and other carbon compounds (see next cycle). Three classes of metal-containing hydrogenases are known: Fe hydrogenases, NiFe hydrogenases and NiFeSe hydrogenases.

The crystal structure of the NiFe hydrogenase from *Desulfovibrio gigas* has been recently solved at 2.85 Å resolution. The crystalline enzyme is in the 'unready' state, a mode which requires a slow (hours) reductive activation. In spite of this, details of the structure allow a reasonable assessment of the nature of the catalytic site and of the mechanism of action. It is believed that hydrogen diffuses to a NiFe active site in the larger subunit (60kDa) where it is heterolytically cleaved to H^+ and H^- (E represents the NiFe centre):

$$E + H_2 \rightarrow EH^- + H^+ \qquad (8.6)$$

The proton is channelled to the outside of the larger subunit. The H^- moiety transfers its electron, *via* two [4Fe4S] and one [3Fe4S] cluster in the smaller subunit (28 kDa), finally to the surface of the smaller subunit where an electron acceptor/donor protein is bound. The reaction is *completely* reversible since the enzyme can catalyse both hydrogen consumption and production. In addition to the crystallographic study, the use of EXAFS, EPR and Mössbauer techniques have contributed to the analysis of the mechanism.

> Q. How would you explain the catalysis of H_2/D_2O exchange by hydrogenases?

> The importance of a metal centre (usual in hydrogenases) for the activation of H_2 is strikingly shown by the surprising discovery of a metal free hydrogenase. This enzyme probably acts by activating the substrate rather than H_2. In this case, a substituted pterin is distorted by the enzyme to form an unstable carbocation which can remove H^- from H_2.

8.4 The methane–carbon dioxide interconversion, E – F

Hydrogen is used by methanogenic bacteria which are responsible for the production of methane in sediments (ponds etc.), oil wells and sludge from sewage plants. The simple reaction (eqn 8.7), which underlies this process,

$$CO_2 + 4H_2 \rightarrow CH_4 + 2H_2O \qquad (8.7)$$

E, is called methanogenesis and as is usually the case, belies the complexity involved in the conversion. A series of enzymatically catalysed reactions proceed through sucessively reduced C_1 functional groups $-CO_2H$, $-CHO$, $-CH_2OH$ and $-CH_3$ bound to various coenzymes including Mo or WFeS and FeS proteins. The final step:

$$CH_3S(CH_2)_2SO_3^- + H_2 \longrightarrow CH_4 + HS(CH_2)_2SO_3^- \quad (8.8)$$

methyl S-coenzyme-M   coenzyme-M

Methyl-S-coenzyme reductase is dimeric and contains a prosthetic group which has an intense yellow colour. The absorption maximum at 430 nm is typical for planar Ni(II) complexes — in this case a Ni(II) tetrapyrrole coenzyme.

requires a nickel tetrapyrrole complex (factor F_{430}) as the prosthetic group in methyl-S-coenzyme-M reductase. The reducing capacity of H_2 in all the steps of methanogenesis is provided by the electrons of eqn 8.5, the equilibrium being established by various hydrogenases. The transmembrane proton gradients generated and the exergonic process associated with eqn 8.8 are used to synthesise ATP. Methanogens produce about 10^9 tons of CH_4 per year, half of which goes into the atmosphere.

The reverse series of reactions, methylotrophy **F**, in which CH_4 is converted into CO_2 proceeds through the same C_1 intermediates as in methanogenesis (eqn 8.9) but the enzymes are quite different.

$$CH_4 \rightarrow CH_3OH \rightarrow HCHO \rightarrow HCO_2H \rightarrow CO_2 \quad (8.9)$$

There exist vast sheets of methane hydrate beneath ocean floors, either frozen or under high pressure. The CH_4 is caged in a sphere of H-bonded H_2O and is difficult to recover. Serious commercial exploitation may be considered, when there is severe depletion of conventional natural gas deposits. Much attention is being focused on finding simple versions of enzymes which will catalyse the conversion of CH_4 to CH_3OH, a fuel which is easier and safer to store and transport.

Each step is catalysed by specific enzymes which require cofactors. In the first step either the iron containing soluble (sMMO) or the copper containing particulate (pMMO) methane monooxygenase is the catalyst. The soluble form has been thoroughly investigated by a number of research groups (Fig. 2.6). In the second step, methanol dehydrogenase catalyses the conversion of methanol into formaldehyde. The enzyme is a quinoprotein (i.e. containing a quinone cofactor) which requires Ca^{2+} or Mg^{2+} ions to help bind the quinone at the active site. The requirements of metal ions for the enzymes in the last two stages are uncertain.

8.5 The carbon cycle, G and H

The interconversion of CO_2 and organic compounds, in the form of carbohydrates, fatty acids and amino acids, represents one of the most fundamental processes for most living material. These take place in respiration and photosynthesis, which are complementary because the O_2 consumed in respiration is supplied by photosynthesis. The major pathway for the breakdown of carbohydrates ends in the electron transport chain with the O_2 molecule as the terminal electron acceptor. In photosynthesis, solar energy is converted into chemical energy by plants, algae and bacteria but not by eukaryotes. Carbohydrates are synthesized from water and carbon dioxide and oxygen is a by-product. Again, a complex electron transfer chain is used following light irradiation. In both respiration and photosynthesis, ATP is generated from ADP and phosphate. Details of these processes have been given in Section 6.4. Plants and cyanobacteria, in using photosynthesis, act on about 10^{11} tons of carbonaceous material per year. Organic compounds and CO_2 can also be interconverted via CO as an intermediate (**H**). The CO/CO_2 interchange is catalysed in both directions by carbon monoxide dehydrogenases. In anaerobic bacteria these complicated enzymes contain Ni and FeS centres, while those from aerobic bacteria contain molybdopterin (Table 6.4).

Earth-orbiting satellites using infrared detectors have shown, surprisingly, that atmospheric CO is present in as large amounts in the Southern as in the Northern Hemisphere, from the burning of tropical rain forests. Plant decay produces CH_4 which is converted by OH radicals and O_2 into CO and CO_2. Hydroxyl radicals are produced by the effect of sunlight on the ozone layer. As CO accumulates, more OH must be used to generate CO_2, leaving CH_4 untouched. The accumulation of these gases (CO_2 and CH_4) contributes to the 'greenhouse effect' and the rise in the mean global air temperature, with potentially devastating effects.

8.6 The oxygen–water link, I

In our aerobic world, as we have just described, the oxidation of compounds by O_2 and the oxidation of H_2O to O_2 are of paramount importance. All animals, plants, yeasts and some bacteria capture atmospheric oxygen and distribute it to cells where it is reduced to water, with difficulty, in the last step of the respiratory chain. Oxygen is used in the biosynthesis of many compounds in metabolic processes and for the conversion of lipid-soluble to water-soluble molecules for excretion. In addition, oxygen is manipulated by hundreds of enzymes, including many metalloenzymes. Cell components can also convert oxygen to partially reduced forms. The half-couples involved and their standard reduction potentials ($E^{\circ\prime}$, volts, at pH = 7, 25 °C) are shown in Chart 8.1.

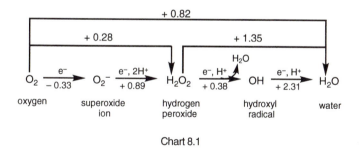

Chart 8.1

Although the conversion of O_2 to H_2O is thermodynamically favourable ($E^{\circ\prime}$ = +0.82 V), multielectron transfer, in this case involving four electrons, is difficult. The stepwise reduction encounters problems at (and only at) the first step, namely the one-electron reduction of O_2 to superoxide ion, O_2^-. As well as a negative reduction potential for the reaction, there are kinetic barriers imposed by spin state restrictions to be overcome. In the ground state oxygen is in the triplet state with two unpaired electrons in the π^*_{2p} orbitals (Fig. 8.4). Most organic substances (X and XO) have singlet ground states (no unpaired electrons) so that the reaction with O_2 can be represented thus:

$$1/2 \;^3O_2 + {}^1X \rightarrow {}^1XO \tag{8.10}$$

$$\uparrow\uparrow \quad \uparrow\downarrow \quad \uparrow\downarrow$$

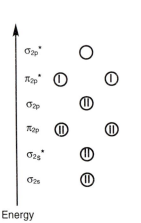

Energy

Fig. 8.4 Molecular orbital energy levels and their filling by the twelve valence electrons of O_2.

Q. In what other ways might the kinetic barrier implicit in eqn 8.10 be overcome and what problems are associated with them?

Q. Cu^{2+} is a good catalyst for the disproportionation of O_2^-. Why would it not be useful for this purpose in the human body?

Since the number of unpaired electrons must be the same before and after an elementary (i.e. one step) reaction, eqn 8.10 cannot proceed rapidly in a single concerted step. Metalloenzymes can aid in the circumvention of these barriers. For example, cytochrome c oxidase binds O_2 to *two paramagnetic* metal ions and in so doing (a) overcomes the spin restrictions and (b) reduces O_2 to O^{2-} (in peroxide) in a two electron step, thus bypassing the unfavourable reduction to O_2^-.

Superoxide ion, O_2^-

The one-electron reduction of O_2 produces the superoxide radical, O_2^-, which having one unpaired electron, will exhibit a characteristic EPR signal. Superoxide is unstable with respect to disproportionation, the rate of which is very pH-dependent. At pHs above 4 the predominant reaction kinetically is:

$$O_2^- + HO_2 \rightarrow O_2 + HO_2^- \tag{8.11}$$

and in (initially) micromolar concentrations of superoxide ion, at physiological pH most (>90%) will have disappeared within about 10 seconds. In even this time, superoxide radical is apparently sufficiently damaging (see below) that biological systems employ an enzyme, superoxide dismutase (SOD) (Table 6.4), to catalyse the disproportionation. There has, however, been some controversy as to whether or not this enzyme is specifically for that physiological function.

Peroxide ion, O_2^{2-}

The π^*_{2p} orbitals are now filled with four electrons and the peroxide ion is diamagnetic. At neutral pHs the ion is diprotonated and H_2O_2 is the dominant form. Like superoxide ion, it too is unstable with respect to disproportionation, although in very pure water its rate of decomposition

$$2H_2O_2 \rightarrow 2H_2O + O_2 \tag{8.12}$$

(eqn 8.12) is very slow and solutions are stable for weeks. Traces of metal ions catalyse the decomposition, a catalytic effect which is used in catalase and to a lesser extent in peroxidase (Table 6.3).

Hydroxyl radical, OH

Further addition of an electron to hydrogen peroxide promotes O–O bond dissociation and the formation of the hydroxyl radical with the unpaired electron on the oxygen atom:

$$H_2O_2 + e^- + H^+ \rightarrow H_2O + OH \tag{8.13}$$

The hydroxyl radical is also formed by the irradiation of water (eqn 8.14) and as with the other oxygen species above, it is very damaging to most biological material (next section). The hydroxyl radical is easily reduced to water, with the aid of a proton (Chart 8.1).

$$4H_2O \rightarrow 2.6e^- + 2.6OH + 0.6H + 2.6H^+ + 0.4H_2 + 0.7H_2O_2 \tag{8.14}$$

Toxicity of oxygen species

High toxicity is associated with O_2^-, O_2^{2-} and OH molecules, and even pure oxygen in high concentrations can cause lung damage. All of these reduced forms of O_2 are produced *in vivo*. The mode of action of O_2^- is still unclear although irreversible damage to joints (arthritis), Alzheimer's disease and even the process of ageing itself has been attributed, at least in part, to free radicals including superoxide. Superoxide and peroxide ions may promote damage because of their ability to produce hydroxyl radicals, eqns 8.15 (Fenton's reaction) and 8.16, or even by reaction of O_2^- with NO (Section 8.9), a known endogenous species (eqn 8.17).

$$Fe(II) + H_2O_2 \rightarrow Fe(III) + OH^- + OH \tag{8.15}$$

$$O_2^- + H_2O_2 \rightarrow O_2 + OH^- + OH \tag{8.16}$$

$$NO + O_2^- \rightarrow ONO_2^- \xrightarrow{H^+} NO_2 + OH \tag{8.17}$$

A hydroxyquinoline derivative (O-Trensox) as the Fe(II) chelate, does not catalyse the damaging Fenton's reaction. The soluble Fe(III) complex can be used to alleviate iron deficiency in plants, while the free ligand has potential as an oral drug to treat iron overload (Section 7.8).

Hydroxyl radical is a powerful and non-selective oxidant which can abstract hydrogen atoms from organic molecules. Like other free radicals it can impair the functions of cellular constituents, lipids or proteins in membranes for example, and DNA in mitochondria and chromosomes. It can modify all DNA bases (unlike O_2^-, H_2O_2 and singlet oxygen) and cause strand scission by degrading the ribose ring. Damage to DNA interferes with the accuracy of transcription leading to faulty gene expression and resulting undesirable mutations in some cases. There are a number of antioxidants, both enzymes and nonenzymes within the cell which control the concentrations of these pernicious oxygen species. These include the CuZn SOD and catalase in the peroxisomes, Mn SOD and peroxidases in the mitochondria, vitamin C (ascorbic acid) and glutathione in the cytosol and vitamins C and E in the lysosomes (Fig. 6.1).

Amyotrophic lateral sclerosis, also called Lou Gehrig's disease after the famous baseball player, arises in about $1/4$ of the victims of the familial form (FALS) from dominantly inherited mutations in SOD1, the gene that encodes human CuZn SOD. Experiments indicate that the FALS-associated mutant protein is a better catalyst of certain oxidation reactions than the wild type, and that this may be the basis of the neuropathalogical changes which occur in FALS.

8.7 Oxygen transport and storage

The reactions of O_2 considered so far all involve its net transformation. However, it is apparent that O_2 must also be carried and stored, virtually unchanged, for use. Finely constituted systems must be able to do this and also to protect the O_2 molecule from its own reactivity. The simple diffusion of O_2 across a cell membrane is adequate for its supply to lower organisms such as small animals and plants. For higher organisms a complex carrier system must be employed, which effectively enhances (by up to 10^2-fold) the concentration of oxygen in water (normally ~millimolar). It is a little surprising that in the natural world there are only three types of respiratory proteins which are responsible for the transport and storage of oxygen (Table 8.1). All of them have been known since the early 1800s. In spite of their

The oxy forms of these proteins continuously self (auto) oxidize to the met form *in vivo*. An enzyme reductase system and reductants in the cell maintain the met content at low levels of only 0.5–1% in normal human blood. Hereditary disorders or exogenous agents can increase the amount of met to dangerous levels (30–50%) resulting in methemoglobinemia. Individuals with hemoglobin M, rather than the normal adult hemoglobin A, have either the α- or β - chain with only Fe(III) present, as a result of an amino acid substitution at or near the heme.

Table 8.1   Naturally occurring carriers of oxygen

Respiratory protein	Source	Characteristics
Hemoglobin 16 kDa	Extensive in animal kingdom and some plants. Human and sperm whale forms have been well investigated.	Scarlet (oxy), purple (deoxy). Largely as tetramer ($\alpha_2\beta_2$ subunits). The subunit is very similar to myoglobin. $Fe^{(III)}-O_2^-$ (or possibly $Fe^{(II)}-O_2$) unit in oxy.
Hemocyanin 60 kDa	In many invertebrate species. Especially examined forms are from octopus and snail (molluscs) and lobster and crab (arthropods).	Blue (oxy), colourless (deoxy). Highly oligomeric (12–150 subunits). $Cu^{(II)}-O_2^{2-}-Cu^{(II)}$ unit in oxy.
Hemerythrin 13.5 kDa	Widespread but limited to several phyla of marine invertebrates. Metalloprotein from sipunculids (peanut marine worms) most examined.	Burgundy (oxy), colourless (deoxy). Usually octameric in coelomic fluid and monomeric in muscle. $Fe^{(III)}XFe^{(III)}-O_2^{2-}$ unit in oxy.

similar names, they are radically different and only hemoglobin contains a porphyrin ring. This group of proteins has been studied more intensely than any other. Most of the techniques of Table 3.1 have been used to characterize the metal sites and the interaction of O_2 with them. There are three common forms for each: deoxy, in which the metals are fully reduced; oxy, formed by the reaction of O_2 with deoxy; and met, in which the metal is oxidized and unreactive with oxygen. When oxygen reacts with the reduced deoxy form, there is a strong driving force which favours oxidation of the metal ion to the met form over reversible oxygen addition to the deoxy form. Irreversible oxidation of the metal is curtailed by its being located in a hydrophobic pocket and by hydrogen bonding at the active site of the respiratory protein. The mode of attachment of oxygen differs in each case (Table 8.1). In general, monomeric forms (termed myo) of the proteins are used in the muscles of the creature for storage of O_2 while its carriers are the polymeric forms. Cooperativity, which we shall discuss shortly, in the uptake of oxygen, is rarely observed with hemerythrin, has been little studied in hemocyanin but is very important and well investigated with hemoglobin.

Myoglobin and hemoglobin

Oxygen is taken up in the lungs, transported in the blood stream in erythrocytes (red blood cells) and deposited in muscle where it is stored and used. The reaction of oxygen with myoglobin (Mb) is, at least superficially (see below), straightforward. The binding conforms to:

$$Mb + O_2 \rightleftharpoons MbO_2 \qquad (8.18)$$

and exhibits the expected hyperbolic relationship between the percentage of saturation and the partial pressure of oxygen (Fig. 8.5). In contrast, the corresponding curve for hemoglobin (Hb) exhibits complex behaviour (sigmoidal) with a pH-dependence (the Bohr effect) not observed with myoglobin. The fraction of oxygen bound by myglobin or hemoglobin (f) vs the partial pressure of O_2 (p) can be expressed by the general relationship:

$$f = Kp^n / 1 + Kp^n \qquad (8.19)$$

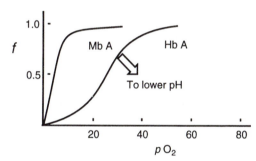

Fig. 8.5 Fraction of saturation (f) vs partial pressure of O_2 (mm) for binding of myoglobin and hemoglobin in whole blood. The partial pressures of O_2 in muscle and lungs are ~40 and ~100 mm, and thus in muscle myoglobin takes O_2 from hemoglobin. Whole blood contains a number of allosteric factors (e.g. bisphosphoglycerate and other organic polyphosphates) which reduce the O_2 binding ability of deoxyhemoglobin. These factors are necessary, otherwise O_2 binding in oxyhemoglobin would be too strong.

where K is the constant for simple oxygen binding to the heme iron and $n = 1$ for myoglobin and $n \sim 2.8$ for hemoglobin. This behaviour indicates that as the heme group in each of the four subunits of hemoglobin binds to a molecule of oxygen, the one or more remaining hemes acquire an increased affinity for O_2. Much time and effort has been spent over many years in attempts to understand the phenomenon of *cooperativity*. The elucidation of the three-dimensional structures of myoglobin, hemoglobin and their derivatives has been basic to this understanding and is a landmark in biochemistry. An approximate, and of necessity, brief probable explanation of cooperativity follows.

The Bohr effect arises from the fact that binding of O_2 by deoxyhemoglobin leads to a release of protons. This means that a decrease in pH will promote O_2 release. Vigorous exercise results in worked muscles which have consumed oxygen and released CO_2. The latter binds to the terminal amino group of hemoglobin to form a carbamate and release a proton. This leads to the desired additional release of O_2.

Q. Construct plots of log $f(1 - f)^{-1}$ vs log p (Hill plots) for $n = 1$, $n = 4$ and $1 < n < 4$. If only $Hb(O_2)_4$ formed what would be the value of n?

The Bohr effect was named after the biologist Christian Bohr, father of the physicist Niels Bohr. A *marked* effect of pH on oxygen uptake has been termed the Root effect. Teleost fish, e.g. tuna or trout, can secrete lactic acid into the blood from a gland attached to the swim bladder. This causes a substantial discharge of O_2 into the bladder and raises the buoyancy of the fish.

Bicarbonate ion greatly supresses the binding of O_2 to crocodile Hb (Fig. 8.6). This means that when crocodiles hold their breath, the bicarbonate that accumulates causes a large fraction of Hb-bound O_2 to be released. This allows crocodiles to stay underwater for an hour or more and drown their prey (if not already dead!). The effect of bicarbonate is small with human Hb, but replacement of twelve amino acids in Hb(A) by site directed mutagenesis (to give Hb Scuba!) reproduced the crocodile effect. The replacements were mainly at the $\alpha_1\beta_2$ (Section 3.4) subunit interface where the bicarbonate binding site is located.

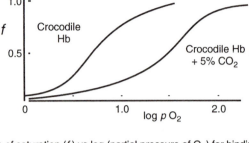

Fig. 8.6 Fraction of saturation (f) vs log (partial pressure of O_2) for binding of recombinant *Crocodylus niloticus* hemoglobin, with and without 5% CO_2 at pH 7.4 and 25 °C. The curve for recombinant HbA (very similar to 'stripped' blood) resembles crocodile Hb, but there is only a small effect of CO_2.

Each subunit of hemoglobin contains an iron protoporphyrin IX centre. In deoxyhemoglobin (and deoxymyoglobin), the Fe(II) is high spin and the fifth coordination site is occupied by an imidazole from the proximal histidine with the sixth ligand probably H_2O. The Fe(II) is too large to fit into the porphyrin ring and is displaced from the plane of the four nitrogens by 0.4–0.5 Å on the histidine side. In oxyhemoglobin, the smaller Fe(III) fits nicely into the ring and in so doing changes the orientation and position of the coordinated histidine. The Fe–N (his) bond, which is about 8° off normal in deoxyhemoglobin, becomes almost perpendicular to the porphyrin plane in oxyhemoglobin. The movement of the Fe–heme–his entity on O_2 binding leads to a reorientation and translation of the $\alpha_2\beta_2$ dimer relative to the $\alpha_1\beta_1$ dimer. This disrupts a series of salt bridges that stabilize the deoxy form (Section 2.4) and leads to enhanced O_2 binding and this continues until all four heme irons are ocupied by oxygen molecules.

In the low temperature X-ray experiments the sample must be frozen carefully to avoid loss of structure. A monochromatic laser beam is used to initiate the Mb–CO bond rupture and synchrotron generated X-rays are used to produce diffraction data. It is hoped that this approach will also be useful for structural characterization of short lived intermediates in enzyme reactions.

High rates of O_2 binding and release are physiologically necessary with all respiratory proteins. Specialized equipment was developed for the measurement of these rates. The invention of the flow method in 1923 for the study of rapid reactions in solution (Table 3.2) stems from the need to measure the rates of the interactions of hemoglobin with O_2 and CO. It is now possible to investigate the reactions of myoglobin with O_2, CO and NO in great detail. For example, irradiation of oxymyoglobin, A, with short, very intense, laser pulses produces transients which are ascribed to B and C in Fig. 8.7. After photodissociation, the ligand remains in the protein for several hundred nanoseconds before returning to the iron in a complex reaction. Until recently only spectral and kinetic evidence for transients existed. However, a crystal structure of irradiated MbCO at < 40 °K visualized the geminate state corresponding to B with the C≡O unattached in a well defined site near the Fe and parallel to the heme plane rather than normal to the ring as when bound to the Fe.

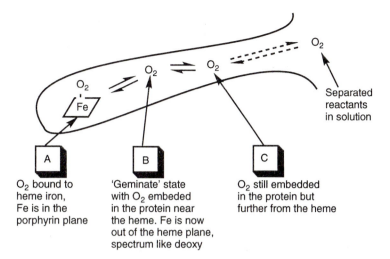

Fig. 8.7 Just one of many pathways for O_2 travel to and from the heme centre and the bulk solution.

The preparation of model metal complexes, particularly to simulate the spectral properties of these respiratory proteins, has been important in understanding them. Indeed, the details of the orientation of the O_2 moiety in the binuclear copper site in oxyhemocyanin was correctly established by using model complexes, before the crystal structure of the protein was solved (Section 3.5). Simulating the function (reversible oxygenation in water at ambient temperature) of the respiratory proteins has not yet been accomplished; however, significant contributions to the understanding of the behaviour of the iron heme proteins have been made by using iron porphyrin complexes of ingenious design in non-aqueous solutions.

8.8 Other important p-block molecules

It is clear from consideration of the cycles in Sections 8.2–8.6 that small molecules containing p-block elements are critically important to the viability of all living systems. In the final sections we will briefly consider a few topics and molecules of groups 15 and 16 which are relevant to inorganic biochemistry.

8.9 Nitric oxide

Nitric oxide (NO) is a colourless reactive gas, first discovered in the early nineteenth century. It has provided one of the biggest recent biological surprises. It has been shown in the past few years that nitric oxide is:

(a) formed endogenously in mammals from L–arginine by a five-electron oxidation (eqn 8.20).

Both reactions are catalysed by nitric oxide synthases. NO has been identified in the human brain and (like CO) detected in exhaled breath.

$$(8.20)$$

The story of NO as a messenger starts with basic research on the network of nerves in the retractor muscle in bull's penis. A number of bodily functions rely on a chemical signal or neurotransmitter released by nerves. What was the chemical? A small organic molecule (e.g. peptide) seemed the logical candidate, but extensive research finally traced it to the gas, NO. The release of NO must be carefully controlled, since it is toxic and unstable. It is interesting that NO is generated in as primitive a species as the horseshoe crab, so it is a system which probably has been in use for at least 500 million years.

Q. What reactions of NO are likely to be important in biological systems?

Q. The human body can produce excessive amounts of NO leading to lethal blood vessel dilation. How might this be treated?

(b) a vital and the first gaseous biological messenger.

It carries out more important functions than virtually any other messenger molecule. It causes muscle cells to dilate and relax partly through the activation of guanylate cyclases (Section 8.10) thus lowering blood pressure. Its formation is regarded as the main determinant of blood pressure in many species, including man. Nitrovasodilators have been known for a long time (see Table 8.2) and it is now thought that they may all act by providing NO. Nitric oxide also participates in the transmission of nerve impulses which is important in a range of activities, including gut contraction and food movement as well as penile erection.

(c) an effective killer of cells (cytotoxic) or blocker of cell multiplication and growth (cytostatic).

It is produced in larger concentrations in (c) than in (b) after activation of macrophages (white blood cells) which destroy bacteria. Its known antibacterial function possibly occurs via the generation of other radicals.

There are at least three distinct NO synthases (NOS), nNOS from neuronal tissue, eNOS from vascular endothial cells and iNOS induced in macrophages. All use a protoporphyrin IX heme and a number of coenzymes. The lack of neuronal NOS genes in mutant male mice does not prevent their survival and breeding but does induce excessive aggression. Maybe nNOS plays a specific role in controlling mammalian social behaviour. There are a variety of inhibitors of nitric oxide synthase action and these may be of value in the treatment of certain clinical conditions arising from irregular production of NO (e.g. restoration of normal blood pressure after endotoxic shock). Not surprisingly, NO was named 'Molecule of the Year (1992)' by the journal Science, and featured in a 1993 BBC TV Horizon program.

8.10 The phosphate ion and its offspring

The phosphate ion and derivatives of it are without doubt among the most important moieties of the p-block in biochemistry. Adenosine triphosphate (ATP) is the most common currency for energy transactions and is probably present in every living organism except viruses. With three negatively charged phosphate groups attached (Fig. 8.8) the molecule is unstable toward hydrolysis and one or both of the P–O–P bonds may be broken by water. The energy thus released can be used to drive other reactions. The hydrolysis of ATP and the reverse reactions (formation) are conveniently written in biochemistry as in eqns 8.21 and 8.22 with no charges shown and with P_i and PP_i representing the protonated forms of phosphate and pyrophosphate ions present at the pH of the reaction. At pH ~ 7 *in vivo*, the reaction shown in eqn 8.21, for example is *mainly* as shown in eqn 8.23.

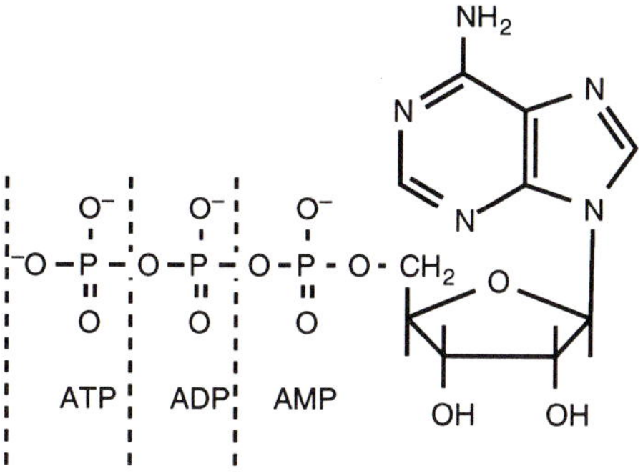

Fig. 8.8 The structures of AMP, ADP and ATP.

Pyrophosphate, $P_2O_7^{4-}$, is an effectiveinhibitor of hydroxyapatite crystal growth which is a cause of tartar on teeth (Table 5.4). Unfortunately, $P_2O_7^{4-}$ is broken down by phosphatases which are present in mouth bacteria. This can be prevented by blocking the metal ion sites in the phosphatases (Table 7.1). A copolymer containing the $(-CH_2CH(OCH_3)CH(CO_2H)CH(CO_2H)-)$ unit is such an inhibitor and is now added to certain toothpastes to prevent tartar formation.

$$ATP + H_2O \rightleftharpoons ADP + P_i \qquad (8.21)$$

$$ATP + H_2O \rightleftharpoons AMP + PP_i \qquad (8.22)$$

$$Mg^{2+}ATP^{4-} + H_2O \rightleftharpoons Mg^{2+}ADP^{3-} + HPO_4^{2-} + H^+ \quad (8.23)$$

ATPases and ATP synthases are enzymes which can catalyse reactions (8.21) or (8.22) in the forward and reverse directions, respectively, depending on the conditions. They can therefore help in pumping H^+ across a membrane or in using H^+ to aid synthesis, mainly in oxidative phosphorylation and photosynthesis (Section 6.4). Another reaction of interest to us is the conversion (eqn 8.24) which is catalysed by guanylate cyclase.

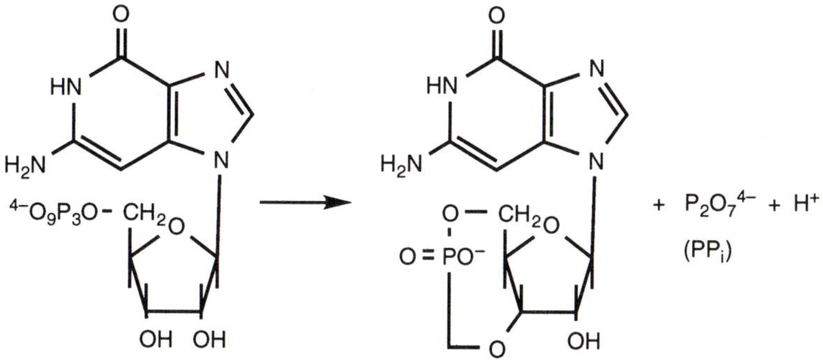

guanosine 5'-triphosphate   cyclic guanosine-3',5'-monophosphate (cGMP)

$$(8.24)$$

The large masses of nutrients, especially phosphates from fertilizers and detergents, discharged into Lake Erie has led to excessive production of algae (eutrophication). Their decay has used up O_2, which is detrimental to fish and has prevented complete breakdown of organic molecules. A concerted effort to reduce discharges of phosphates into Lake Erie is leading to slow recovery of a dying lake.

The secondary messenger cGMP decreases the amount of free Ca^{2+} ion in the muscle cell, which leads to relaxation of the vascular smooth muscle (this muscle surrounds and controls the diameter of blood vessels). Nitric oxide and carbon monoxide bind to the heme portion of guanylate cyclase, produce a conformational change and activate the enzyme.

8.11 Ubiquitous water

Another environmentally friendly solvent is CO_2 which is a liquid above 31° C and 72.8 atmospheres. It has been used in the production of decaffeinated coffee. Recently water-in-CO_2 microemulsions, induced by perfluoropoly ether surfactants, have been found to dissolve proteins. This has unique potential applications.

Water has been observed in the spectra of red giant stars and, very recently, intact within sunspots, where temperatures are around 3000 °C. Its importance on earth is self evident. Were it not for hydrogen bonding, water would be a gas like H_2S at ambient temperature and pressure. It is, of course, a liquid and forms the medium in which most biological reactions take place. When it freezes at 0 °C, the solid ice floats, and this is also a result of extensive hydrogen bonding which produces a more open structure. It is very unusual for the solid form of an element or molecule to be less dense than the liquid form. In water this effect is vital, since otherwise ice would sink to the bottoms of lakes, ponds etc., remain unthawed and build up to the detriment of marine life. A hope for the future is the use of special forms of water to dissolve or convert hazardous wastes. At high pressures, the boiling point of water is substantially raised (374 °C at >200 atmospheres). Under these conditions it is a supercritical fluid and has the properties both of a gas and a liquid. In this state, it has the ability to dissolve oils or sewage sludge and even slowly corrodes gold. Ultrasound creates tiny bubbles for short times which burst at high temperature and pressure and convert some H_2O into OH radicals. These can be used to transform chlorofluorocarbons into CO_2 for easy disposal.

In fish, antifreeze proteins bind via many H bonds to ice surfaces and inhibit the growth of ice crystals, thus protecting the fish from freezing. Certain antifreeze glycoproteins also allow some Antarctic fish to live at temperatures below 0 °C.

It probably comes as a surprise that some organisms, as well as their associated enzymes, termed extremozymes can tolerate extreme conditions (high temperatures, high salt or low pH). Hyperthermophilic organisms can grow optimally at 100 °C or above, are usually prokaryotes and similar to bacteria, and are isolated from geothermal environments. They have attracted interest because the conditions in which they grow mimic those of prehistory on earth. In addition, they may have practical applications. Xylose or glucose isomerase (Section 7.2) from *Thermotoga maritima* is optimally active at 100 °C, at which temperature there is a higher (desired) fructose concentration in the equilibrium mixture than at the lower temperatures usually employed.

8.12 The sulfur cycle

Sulfur has a prominant role in biology and the sulfur cycle consists of a set of processes in which it circulates within the biosphere. In plants, SO_4^{2-} is the source of cysteine, which animals must obtain from plants. A group of unrelated bacteria containing metalloenzymes handle the transformations shown in eqn 8.25, some of which are reversible:

$$S^{2-} — H_2S — SO_3^{2-} — SO_4^{2-} — \text{organic S}$$

with S bridging from S^{2-} to SO_4^{2-}, and H_2S bridging to organic S.

$$(8.25)$$

Assimilatory sulfite reductases take SO_3^{2-} to HS^- for use in biosynthesis:

$$SO_3^{2-} + 6e^- + 7H^+ \rightarrow HS^- + 3H_2O \qquad (8.26)$$

The electrons are supplied by NADPH. The SO_3^{2-} is bound to a siroheme (type d) iron which in turn is linked by cys–S to an Fe_4S_4 cluster. The enzyme from *E. coli* is also an effective $6e^-$ assimilatory nitrite reductase (eqn 8.2).

In photosynthesis green and purple bacteria use H_2S from dead organic matter rather than H_2O, which is used by green plants (eqn 8.27):

$$6CO_2 + 12H_2S \rightarrow C_6H_{12}O_6 + 12S + 6H_2O \qquad (8.27)$$

Hydrogen sulfide is a deadly poison. The Black Sea is the world's largest single reservoir of H_2S. About 150 metres below its surface there is no life, but this condition has existed for thousands of years. The influx of organic material from the rivers flowing into the Black Sea used up the oxygen, so that bacteria had to feed on SO_4^{2-} and thus produced H_2S. This situation has now been aggravated by a large influx of modern pollutants and the result has been a sharp decrease in the types of fish in the sea and fish catches.

8.13 Health and the p-block

Table 8.2 shows some common pharmaceuticals which contain p-block elements. Where applicable, doses are relatively large unless the material is toxic as well as beneficial. The substances can be administered as tablets, capsules or as a suspension or solution.

A dilute solution of Nipride is passed into the blood stream during major surgeries, especially transplants, to lower blood pressure. At completion, infusion is stopped and blood pressure returns to normal without overshoot.

Table 8.2   Inorganic pharmaceuticals

Compound	One trade name	Value
Na perborate	Bocasan	Antiseptic; mouthwash
$Al(OH)_3$	Anhydrol Forte	Antiperspirant
$Mg_3(OH)_2Si_4O_{10}$	Talcum powder	Mildly antiseptic cleansing powder
SnF_2	Fluoristan	Caries-preventing toothpastes
N_2O		Anesthetic
$Na_2Fe(CN)_5NO$	Nipride	Vasodilator (Section 8.9)
As_2O_3		Skin and blood disorders (rarely used)
Na stibogluconate (as a paste)	Pentostam	Treatment of skin infections e.g. for leishmaniasis
$K_3Bi(citrate)_2$ (plus antibiotics)	De-nol / Tritec	Promotes healing of peptic ulcers
SeS_2	Selsun	Antidandruff agent
Na hypochlorite and chlorine 'active' compounds		Disinfectants
NaBr		Sedative
Iodine		Antiseptic

Certain boron compounds can localize in tumors. Neutron irradiation at that point can produce lethal doses of radiation *in situ* (neutron capture therapy):

$$^{10}_{5}B + ^{1}_{0}n \rightarrow ^{11}_{5}B$$

$$^{11}_{5}B \rightarrow ^{4}_{2}He + ^{7}_{3}Li + 2.8 \text{ Mev}$$

8.14 The Future

Predicting the important future developments in bioinorganic chemistry is more difficult than picking the winner of a horserace. The winner in science may not even have been in the starting lineup! For example, who would have guessed that a toxic gas (nitric oxide) would turn out to be an important chemical messenger, and who would have imagined that a simple platinum complex would be a clinically useful anticancer agent?

We can however, with some confidence, suggest areas and problems with which future researchers will be concerned. There are still large gaps in our understanding of the uptake, storage and roles of the trace elements in the human body and their chemical speciation under cellular conditions. Iron especially appears to be an important mediator of oxidative damage *in vivo*. How does this occur? Only quite recently was the reason for the relatively high abundance of Zn^{2+} ion in living systems clarified by the discovery of its part in gene regulation. The importance of other metal ions in this area is also becoming apparent.

The combination of intense synchrotron X-rays with sophisticated computers allow the crystal structures of the most complicated metalloproteins to be solved. Even more important will be efforts to determine the structures of intermediates in enzyme catalysed reactions and thus begin to build up a moving picture of enzyme action. This will be achieved by slowing down the formation and decomposition of intermediates by lowering the temperature or by using genetically engineered mutants in which there is a change in the rate determining step in the process under investigation. Intermediates will then be stable for sufficient time for a crystal structure to be determined.

The value of inorganic compounds in the treatment of disease has so far been mainly a story of serendipity (e.g. Li and Pt compounds). A concerted effort is needed to understand the principles underlying their value and to exploit these to develop the field. One area where rewards for sucess will be high is in the understanding of biomineralization. Already, the manufacture and use of artificial bone is promising. Mixtures of hydroxyapatite and polyethylene are accepted by natural bone and may be used as implants. Some inorganic mixtures can be moulded into any shape.

It should be clear from the necessarily short survey in this book, that a sizable chunk of the periodic table is involved in the biological functions of all organisms. The chemistry of these elements and their compounds have been thoroughly explored, so that there is a firm basis for understanding their roles in biology. They may be involved only in trace amounts, and as simple ions, metal complexes or as components of large biological molecules. The relationship of these elements to all facets of biochemistry and molecular biology, pharmacology, clinical and environmental chemistry will be examined in an exciting, assured future.

Further reading

General *Biological chemistry of the elements.* da Silva, J. J. R. F. and Williams, R. J. P. (1991). Oxford University Press.
Inorganic biochemistry. Cowan, J. A. (1993). VCH, Weinheim.
Principles of bioinorganic chemistry. Lippard, S. J. and Berg, J. M. (1994). University Science Books, Mill Valley, California.
Bioinorganic chemistry. Kaim, W. and Schwederski, B. (1994). Wiley, NY.
Biochemistry. Stryer, L. (1995). Freeman, New York.
Biochemistry. Voet, D. and Voet, J. G. (1995). Wiley, New York.
Chapter 2 Structural aspects of metal liganding to functional groups in proteins. Glusker, J. P. (1991). *Advances in Protein Chemistry,* **42**, 1-76.
Biocoordination chemistry. Fenton, D. E. (1995). Oxford University Press.
Chapter 3 Metallobiochemistry. *Methods in Enzymology,* Volumes **158A**, **205**, **226 and 227** (ed. J. F. Riordan and B. L. Vallee) Manipulations and spectroscopic characterization of metalloproteins.
Chapter 4 Lectin–carbohydrate complexes of plants and animals. Sharon, N. (1993). *Trends in Biochemical Sciences*, **18**, 221–30.
Zinc fingers. Rhodes, D. and Klug, A. (1993). *Scientific American*, **268**, (2), 32–9.
Chapter 5 Calcium in biological systems. Forsén, S. and Kördel, J. (1994). In *Bioinorganic chemistry*. (ed. I. Bertini, H. B. Gray, S. J. Lippard and J. S. Valentine). University Science Books, Mill Valley, California, pp. 107–66. This is one of nine authoritative articles.
Chapter 6 Why do we age? Rusting, R.L. (1992). *Scientific American*, **267**, (12), 131-41.
Iron–sulfur proteins. (ed. R. Cammack and A. G. Sykes) (1992). *Advances in Inorganic Chemistry*, Vol. 38.
Electron transfers in chemistry and biology. Marcus, R. A. and Sutin, N. (1985). *Biochimica et Biophysica Acta*, **811**, 265–322.
Electron tunneling in proteins. Langen, R., Chang, I.J., Germenas, J.P., Richards, J.H., Winkler, J.R. and Gray, H.B. (1995) *Science*, **268**, 1733–35.
Chapter 7 *Trace element medicine and chelation therapy.* Taylor, D. M. and Williams, D.R. (1995). Royal Society of Chemistry, Cambridge.
Transition metals in control of gene expression. Halloran, T.V. (1993). *Science*, **261**, 715-24.
Clinical aspects of platinum anticancer drugs. Rosenberg, B. (1980). *Metal Ions in Biological Systems*, **11**, Chapter 3.
Chapter 8 Bioinorganic reactions of nitric oxide underly diverse roles in living systems. Rawls, R. (1996). *Chemical and Engineering News*, May 6, 38-42.
Inorganic chemistry and drug design. Sadler, P.J. (1991). *Advances in Inorganic Chemistry*, **36**, 1-48 (also Chapters 5 and 7).

Answers to questions

Chapter 1 p.1 Seawater (H, O, Cl, Br, S), minerals (Na, K, Mg, Ca, Ti, Fe, Al, C, Si, O, S) atmosphere (O, N, S) and organic material (C).

p.2 New perspectives on the essential trace elements, Frieden, E. (1985). *Journal of Chemical Education* **62**, 917–23.

p.3 First row availability and ions undergo reactions more rapidly (enzymes).

Chapter 2 p.7 δ- or ε-N coordination. Both Ns bridge (as N and N⁻). Carboxylate can bind by one or two Os to one metal or bridge two metals.

p.7 Broad efforts focus on synthesis and characterization of porphyrins, Borman, S. (1995). *Chemical and Engineering News*, June 26, 30–1.

p.12 NH_2, CON^-, CON^- and imid N form planar Cu(II) complex.

Chapter 3 p.16 Voltammetric studies of the interaction of tris(1,10-phenanthroline)cobalt(III) with DNA, Carter, M. T. and Bard, A. J. (1987). *J. Amer. Chem. Soc.* **109**, 7528–30.

p.18 Direct exchange of metal ions. Biosynthesis in absence of native metal ion and presence of replacement ion.

p.18 Tb^{3+} for Ca^{2+} (fluorescence); Mn^{2+} for Mg^{2+} (EPR).

p.18 Higher coordination (≥ 6) for Ca^{2+} does not allow substrate attachment. Zn^{2+} simulates Mg^{2+} (crystal structure on Zn^{2+} form).

p.19 K_i values for Cu^{2+}, Ni^{2+}, Zn^{2+} (21, 49 and 128 µM) reflect decreasing stability of metal complexes.

p.20 Engineered metal-binding proteins: purification to protein folding, Arnold, F. H. and Haymore, B.L. (1991). *Science* **252**, 1796–7.

p.20 Conformational changes may vitiate conclusions. Only naturally occuring amino acids can be interchanged.

Chapter 4 p.21 *Local* positive charges (from lysine and binding site) control ionic strength effects.

p.22 O_2^- ion would bind too strongly at channel mouth.

Chapter 5 p.30 K^+ to interact with negatively charged biomolecules; Cl^- to maintain electroneutrality.

p.35 $[CaL^{2+}][Ca^{2+}]^{-1}[L]^{-1} \sim 10^6\,M^{-1}$ means $[CaL^{2+}][L]^{-1} \sim 10^{-1}$ at $[Ca^{2+}]$ ~ 0.1 µM and ~ 10 at $[Ca^{2+}] \sim 10$µM. L = ligand.

p.35 Much stronger and more rapid binding by Ca^{2+}. Higher and more distortable coordination.

Chapter 6 p.40 Selectivity to size of metal ion. Thermally (in sediments) and kinetically stable. Form radicals, useful in energy transformations.

p.44 Antiferromagnetic coupling of unpaired electrons on Cu(II) and free radical.

p.46 Direct formation of Fe(IV)O and oxidation of substrate.

p.47 Resemble those of myoglobin.

p.48 Activity unchanged. Thermal stability decreased. Dialysis at pH ~ 4 removes less firmly bound Zn^{2+}.

p.49 Large rapid release of energy would denature proteins etc. (observe effect of plunging smouldering glucose into O_2).

p.51 Leads to phosphorylation of ADP in both cases.

p.53 To transform one $2e^-$ change into two $1e^-$ changes.

p.53 Mg^{2+} fits ring nicely, correct charge and is redox inert so that undesirable reactions are avoided.

Chapter 7 p.58 CO_2 required to form carbamate from ε-amino residue of lys to bridge 2Nis.

p.63 Acetidine N, secondary amino, both terminal carboxyl (plane) and hydroxyl O and intermediate carboxyl (axial) promote octahedral coordination around Fe(III). There are four chiral centres.

p.66 Mg^{2+} binds phosphate in DNA which increases helix stability and melting temperature. Cu^{2+} prefers base binding which destabilizes helix and results in a decrease in melting temperature.

p.66 Natural abundance high. Redox inactive so unlikely to cause DNA damage. Unlikely to coordinate to base in DNA and interfere with protein binding.

p.69 Administering calcium salt (gluconate) or using $CaEDTA^{2-}$.

p.71 Palladium(II) much more labile than Pt(II) and complexes dissociate readily. Trans complex cannot form intrastrand crosslink.

Chapter 8 p.75 Combination of $E + H_2 \leftrightarrow EH^- + H^+$; $EH^- + D^+ \leftrightarrow HD + E$ and $H^+ + D_2O \leftrightarrow HDO + D^+$ give $H_2 + D_2O \leftrightarrow HD + HDO$.

p.78 Two suggestions: (a) $\frac{1}{2}\,^3O_2 + {}^1X \rightarrow {}^3XO$, ${}^3XO \rightarrow {}^1XO$; (b) ${}^3O_2 \rightarrow {}^1O_2$, ${}^1O_2 + {}^1X \rightarrow {}^1XO$. Energetically unfavourable.

p.78 Toxic and coordinates to cell constituents.

p.81 If only $Hb(O_2)_4$ formed, $n = 4$.

p.84 Reactions with $O_2 (\rightarrow NO_2)$, $O_2^- (\rightarrow ONO_2^-$, potent oxidant), RSH ($\rightarrow$ RSNO, s-nitrosothiol) and heme ($\rightarrow$ nitric oxide adduct).

p.84 Use NO synthase inhibitor, e.g. L-thiocitrulline.

Index